AF577188

Biodiversity for Sustainable Development

Biodiversity for Sustainable Development

Edited by

Azmal Hussain

2012

Icfai Books

The Icfai University Press

Biodiversity for Sustainable Development

Editor: Azmal Hussain

First Edition: 2012
Printed in India

Published by

This book is published by IUP.
University Campus, Agartala-Simna Road,
P.O. Kamalghat Sadar, Agartala – 799210, Tripura (West)
E-mail: info@iupindia.org
Website: www.books.iupindia.org

ISBN: 978-81-314-2707-1

Contents

Overview

Biodiversity essentially refers to all aspects of variability evident in the living world. It encompasses diversity within and between individuals, populations, species, communities and ecosystems. Biodiversity trends are mainly evaluated in terms of declining populations and species, either individually or collectively. The challenges to biodiversity have been explicit from the facts and figures provided by different agencies concerned. Experts lament that many species are no longer found throughout their former ranges, and may only occur in reduced numbers. Many high value fish have shown considerable reductions, mainly owing to overexploitation. Conservation and management of biodiversity is a priority concern in environmental as well as development policies universally. This is due to the increasingly growing need for maintaining ecological balance throughout the world coupled with the prospects for return from the global biodiversity reserve. Besides, human exploitation beyond a concerning limit may pose challenges to the global wealth of biodiversity every now and then. Against this backdrop, the book attempts to capture some of the issues pertaining to biodiversity conservation and management. It also highlights the existing and potential role of society, including different institutional arrangements in the context of conservation and management of biodiversity. The book emphasizes on these aspects through various country experiences.

The book has been divided into two sections. The first section deals with an overview of various biodiversity issues and concerns. The second section analyzes the status and initiatives in different geographical settings through country experiences.

Section I: Understanding Biodiversity

"Understanding Biodiversity Issues: A Global Perspective" by *Azmal Hussain* is the introductory article of the section. The article starts with explaining the basic concepts of biodiversity. Biodiversity, the variety of living entities, which incorporates plants, animals, micro organisms along with the entire range of genetic information they contain as well as the ecosystems formed by them. For insuring biodiversity in marine as well as terrestrial systems, we need to identify the fact that the entire natural resource base, people's economic development, our basic needs like food, medicine, clothing, the air we need for breathing and, to top it all, our very existence on earth depends on the life surrounding us. Therefore, in preserving biodiversity, we are insuring nothing but our own continued survival and prosperity. There are multifarious factors posing threats to the world's biodiversity. The factors threatening biodiversity in estuaries or in the oceans are usually the same as those affecting biodiversity in terrestrial systems. Alien species introduction, physical change in habitat areas, overexploitation and alteration in atmospheric composition all pose threats to biodiversity. If we choose locations with no consideration to the intra-community variations, area-based thresholds may only be able to protect a small share of the total diversity. In this case, a coarse-filter approach may be applied for identifying gaps existing in a plant community's current reserve system by considering current climatic conditions and the scenario of global climate change as the bases. As more and more regulatory efforts are made concerning access to genetic resources as well as to establish comprehensive benefit-sharing mechanisms, various difficulties are becoming increasingly clear facing in the medium term. Some seemingly inevitable changes may require the scientific community to adapt to the rules set by the Convention on Biological Diversity (CBD). However, some experts feel that more harmonious laws and policies can be facilitated.

The second article ***Policy Brief*** **– "Preserving Biodiversity and Promoting Biosafety"** from OECD opines that governments should aim at maintaining biological variety as well as encouraging economic growth. Once governments decide to act to protect biodiversity, they need to choose policies that reflect the relative value people place on it. Incentive measures are the basis of a market approach to biodiversity management that help reconcile differences between the value of biodiversity-related resources to individuals and the value of biodiversity to society as a whole. Governments have two important roles in supporting markets for biodiversity-related resources. These are: establishing the right framework conditions for private and public operators to supply biodiversity-related resources efficiently to users; and applying the right policy instruments to ensure that public biodiversity-related goods and services are provided in the most efficient and effective manner. In fine, products of modern biotechnology might also affect biodiversity, and thus raise environmental safety issues.

The third article **"Biodiversity Loss Threatens Human Well-Being"** by *Sandra Díaz, Joseph Fargione, F Stuart Chapin III* and *David Tilman* narrates a synthesis of the most crucial messages emerging from the latest scientific literature and international assessments on the role of biodiversity in ecosystem services and human well-being. Many activities indispensable for human subsistence lead to biodiversity loss, and this trend is likely to continue in the future. Most of the concrete actions to slow down biodiversity loss fall under the domain of policy-making by governments and civil society. The environmental changes have large impacts on ecosystem processes, and thus human well being. Finally, the authors say that in order to assist policy decisions and negotiation among different local, national and international stakeholders, considerable advance is needed in the evaluation and accounting of ecosystem services.

In the fourth article "**EPOC *High-Level Special Session on the Costs of Inaction* – The Costs of Inaction with Respect to Biodiversity Loss – *Background Paper*"**, *Geoffrey Heal* provides an explanation as to why biodiversity loss matters. Policy inaction augments the total loss of biodiversity and the services that are dependent on it. The uncertainty, coupled with the irreversibility of biodiversity loss and the possibility of learning more about its value, implies a real option value to the conservation of biodiversity and the rationale for a precautionary approach. While the overall cost of biodiversity loss is unknown, some parts of this cost can be estimated. These are the cost of lost carbon storage, the cost of tourism business and that of reduced watershed protection. The author concludes that though very partial, the costs under different heads amount to tens of billions of dollars.

The fifth article "**The Structure of Biodiversity – Insights from Molecular Phylogeography**" by *Godfrey M Hewitt* explains the emergence and development of molecular phylogeography and its implication in conservation of biodiversity. DNA techniques, analytical methods and palaeoclimatic assessments are encompassed under phylogeography, and such techniques are greatly advancing our knowledge of the global distribution of genetic diversity. The advance in DNA technology is producing a wealth of data for individuals, populations and species, and there are concomitant developments in analytical methods facilitated by access to increasingly powerful computers. The author concludes that understanding of the distribution of biodiversity through molecular phylogeography carries serious implications for the theory and practice of conservation.

The sixth article "**Design and Applicability of DNA Arrays and DNA Barcodes in Biodiversity Monitoring**" by *Mehrdad Hajibabaei, Gregory A C Singer, Elizabeth L Clare* and *Paul D N Hebert* highlights a comparison of two platforms for identification of mammalian biodiversity. The rapid and accurate identification of species is a critical component of large-scale biodiversity monitoring programs. DNA arrays and DNA barcodes are two molecular approaches that have garnered much attention. The analysis shows that DNA arrays and DNA barcodes are valuable molecular methods for biodiversity monitoring programs. Both approaches are found to be capable of discriminating among mammalian species.

The seventh article of the section **"Whither 'Community-Based' Conservation?"** by *Chetan Kumar* focuses on formal approaches in community-based conservation of biodiversity. The community-based conservation approach emerged after the Western model based on separation of nature from culture and people did not succeed in protecting biodiversity loss around the world. The paradigm shift to community-based conservation arose from the recognition that in many parts of the world, conservation was unattainable without the support of the people living in the proximity of the parks and sanctuaries. A dominant argument for community-based conservation emphasizes on seeing conservation and development as co-dependent. The author finally says that to be effective, programs for protecting biodiversity must use a wide range of approaches, which certainly starts with effective participation at all stages of preparation and implementation, from a full range of stakeholders to reconcile the interest of different groups.

The eighth article **"EGenBio: A Data Management System for Evolutionary Genomics and Biodiversity"** by *Laila A Nahum, Matthew T Reynolds, Zhengyuan O Wang, Jeremiah J Faith, Rahul Jonna, Zhi J Jiang, Thomas J Meyer* and *David D Pollock* highlights the EGenBio, one of the most recently emerged strategies designed to serve as a platform for tools and resources to case analysis of biodiversity combines with allied aspects. The system involves manipulation and filtering of large numbers of sequence alignments and phylogenetic trees. The biodiversity division involved in the design took for analyzing individual sequences or sequence alignments. The tools under biodiversity division perform varied functions such as listing of species, searching species by taxonomic group or genome identifier, displaying mitochondrial gene order for specified taxa and (4) Providing information about the biodiversity databases.

Section II: Country Perspectives

The second section starts with the ninth article **"Gender, Local Knowledge, and Lessons Learnt in Documenting and Conserving Agrobiodiversity"**, written by *Yianna Lambrou* and *Regina Laub,* which explores the linkages among gender, local knowledge systems and agro-biodiversity for food security by using the impact assessment of LinKS, a project initiated by Food and Agriculture Organization, in four countries – Mozambique, Swaziland, Zimbabwe and Tanzania. The project aimed at raising awareness on how rural people use and manage agro-biodiversity, and promoting the importance of local knowledge for food security and sustainable agro-biodiversity at local, institutional and policy levels. The project involved a diverse range of stakeholders to strengthen their ability to recognize and value farmer's knowledge and to use gender-sensitive and participatory approach in their work. A crucial learning from the project is that farmers hold very specific local knowledge about the plants and animals they manage. The authors conclude that local knowledge, gender and agro-biodiversity are so much closely inter-related that if one of the three is threatened, the risk of losing biodiversity increases.

The tenth article **"Indigenous Knowledge and Biodiversity Conservation and Management in Ghana"** by *Luc Hens* deals with an analysis of a series of biodiversity-related areas in Ghana, including ecosystem, water and soil management, farming, fishing and hunting practices and the collection of herbal medicines. The analysis shows that traditional water, soil and ecosystem management, prohibitions on fishing and hunting and the ecological value of sanctuaries provide keys to involve traditional knowledge and practice in a more explicit way than before in biodiversity conservation and management. The author argues that maintenance of rules based on tradition is stronger and more community-owned than government rules, which offers an opportunity to involve people in biodiversity conservation. But a

country in fast development transition like Ghana is subject to skepticism about the way traditional knowledge is used and applied. While comparing the potential of biodiversity conservation by indigenous knowledge, the author points to a twofold conclusion: attention should be given to document and inventory traditional knowledge on biodiversity as the available knowledge is at risk of disappearing fast and education should give attention to the values and limitations of indigenous knowledge so that it can obtain a fair place in development paths of countries like Ghana.

The eleventh article "**Nature Makes them Lazy: Contested Perceptions of Place and Knowledge in the Lower Amazon Floodplain of Brazil**" by *Mark Harris* attempts to show the lower Amazonian floodplain in Brazil as a multilayered place made of human labour as well as the labour and its movements. The author says that without taking into account practical knowledge of small-scale farmers, agro-biodiversity and conservation is undermined. In situ maintenance of agro-biodiversity, an analysis of local knowledge management techniques has been cited as an important practice. The knowledge of indigenous people is at times directed at encouraging modern practices and high-yielding varieties of certain species. Until recently, the floodplain of the main Amazon river attracted little attention, perhaps owing to its scanty forest biodiversity and its risky nature caused by the uncertainty of the river's seasonal changes.

The next article "**River Rehabilitation for Conservation of Fish Biodiversity in Monsoonal Asia**" by *David Dudgeon* describes the status, challenges and opportunities in conserving fish biodiversity in Monsoonal Asia, where freshwater biodiversity as well as ecosystems are seriously jeopardized by human activities. The authors point out that management is especially problematic for large rivers in the region crossing several provincial boundaries and falling within the responsibility of many local authorities. After a detailed analysis of various interventions, it is seen that the conflict between economic development and conservation plus the overlapping of sometimes contradictory responsibilities of government authorities have impeded actions that are needed to preserve biodiversity. The activities of non-governmental environmental groups indicate growing societal concern over the management of freshwater resources parallel to longer-established citizen's movements. An overriding priority identified by the authors is to convey the fact that holistic river management and the restoration of the integrity of riverine ecosystems in Asia should be done through the provision of fisheries resources and clean water.

In the thirteenth article "**Linking Biodiversity Conservation and Livelihoods in India**" the authors *Kartik Shanker, Ankila Hiremath* and *Kamal Bawa* speak of biodiversity conservation in India mainly to curtail the rapid loss of the country's biodiversity. The current activities emphasize on inter-disciplinary approaches that are specifically used to: (1) generate knowledge that fosters conservation and judicious management of biodiversity (2) provide the best scientific information to policymakers (3) design management systems that emphasize decentralization, fairness and equity in the use of resources by civil society (4) organize and disseminate information for conservation and sustainable use of biodiversity and (5) train a new generation of leaders to meet current challenge in biodiversity conservation and environmental protection. In the North and Northeast India, the programs are directed toward conservation of the eastern Himalayan region. The inter-related issues of livelihood enhancement and biodiversity conservation have been a major focus in the Western Ghats, one of the biodiversity hotspots. Different centers are engaged in education and outreach activities designed to build the capacity of academic, governmental and non-governmental organizations to meet the growing list of challenges to biodiversity.

The fourteenth article "**Monitoring Biodiversity of Select Restoration Sites in New Zealand**" by *Jesse Bishop, Russell G Congalton* and *Mimi L Becker* deals with biodiversity issues in New Zealand, wherein the government has recognized and taken steps to halt the loss of native biodiversity. As a result of millions of years of isolated evolution, a majority of the flora and fauna found in

New Zealand are unique in the world. But the measurement developed by the New Zealand Department of Conservation shows a steady decline in biodiversity since the time of settlement. Much of New Zealand is now covered by non-native plants and animals, whereupon recently there were increased efforts to mitigate these effects through a process of ecosystem restoration. While the initial restoration efforts were focused on offshore islands, current efforts extend to mainland islands, restoration sites within a larger landscape that are isolated by a natural or constructed barrier.

The fifteenth article, "**Biodiversity Conservation, Sustainable Development and the US Man and the Biosphere Program: Past Contributions and Future Directions**" by *P N Manley* and *D C Hayes,* deals with the US Man and Biosphere (MAB) program that forms part of the UNESCO's MAB Program. The MAB Program of the US is in a period of reflection and revitalization as it nears 30 years of commitment and contribution to biological diversity conservation and sustainable development. Major accomplishments of the MAB program in the US have been through many different institutions such as the Information Centre for the Environment's (ICE) development of a biodiversity database that serves to document species occurrences for protected areas; the development of the monitoring and assessment of biodiversity program research and education activities; and promotion of environmental health and stewardship in natural and cultural ecosystems.

The sixteenth and last article "**Climate Change and Biodiversity in Europe**" by *Hannah Reid* initially explains how climate change is affecting European biodiversity, and subsequently describes the strategies to mitigate such changes. Strategies to mitigate or adapt to climatic change also impact biodiversity. The author stresses the need for more integrated policy responses at different levels. Nationally, activities that meet biodiversity and climate change objectives need promoting and mainstreaming into various sectoral areas of policy-making. Reducing energy consumption, increasing energy efficiency and promoting renewable energy technologies are cited as priorities.

Section I

Understanding Biodiversity

1

Understanding Biodiversity Issues: A Global Perspective

Azmal Hussain

Biodiversity is often used as a measure of the condition of living systems. In some regions biodiversity is consistently rich, while in some others fewer species are found. The last century has increasingly witnessed erosion of global biodiversity owing to species extinctions that occur mostly because of human activities, particularly destruction of habitats for plants and animals. The ever-increasing extinction rates are being triggered by continuous human consumption. Another potent threat to global biodiversity is the widespread introduction of numerous exotic species by humans. Within the broad embodiment of biodiversity, an array of objective measures has been created to measure biodiversity empirically, each relating itself to a particular application of the data. This article attempts to provide an overview of some biodiversity issues having contemporary relevance.

1. Understanding Biodiversity

In simple words, biodiversity may be defined as the variety of living entities, which includes plants, animals, and micro organisms along with the entire range of genetic information they contain as well as the ecosystems formed by them. Biodiversity is often used as a measure of the condition of living systems. Before coining of the term biodiversity, the term "natural diversity" was used in a study titled "The Preservation of Natural Diversity" by The Nature Conservancy in 1975. Even prior to that, scientists like R E Jenkins and T Lovejoy while dealing with conservation issues used the term "biological diversity". This might be followed by the coining of the term biodiversity in mid 1980s by W G Rosen who did so while arranging the "National Forum on Biological Diversity (Wilson and Francis, 1988). In 1992, the United Nations Earth Summit, which took place in Rio de Janeiro offered arguably the most comprehensive definition of "biodiversity" as "the variability among living organisms from all sources, including, inter alia, terrestrial, marine, and other aquatic ecosystems, and the ecological complexes of which they are part: this includes diversity within species, between species and of ecosystems". In its own right, this has been considered as closest to a legally accepted definition, by virtue of its adoption by the United Nations Convention on Biological Diversity.

In its most common form of practice, we explore biodiversity at the following three levels that work together for creating the complexity of the living world.

Genetic Diversity

This is the genetic variety of living organisms within a species. Individuals having their own specific genetic composition form a species. Therefore, different populations may be included in a species, and each of them can have different genetic compositions. It is essential to conserve various populations of a species so as to conserve genetic diversity. Genes are the fundamental units of all forms of life on earth. The differences as well as similarities between living organisms are also governed by genes. Geneticists consider biodiversity as the diversity of organisms as well as their genes. As genes are the basic units of natural selection, genetic diversity is often designated by them as the real biodiversity (Wilson, 1988).

Species Diversity

This refers to the variety of species prevailing within a particular habitat or region. Habitats like rainforests and coral reefs comprise of many species. Others like salt flats or polluted streams have fewer species. Species are normally grouped into different families based on some shared characteristics.

Ecosystem Diversity

Ecosystem diversity is defined as the variety of ecosystems existing in a given geographical area. An ecosystem comprises a community of organisms along with their physical environment and their interaction. It can contain a small area like a pond, or may encompass a large area, for instance a whole forest. To an ecologist, biodiversity also refers to the diversity of enduring interactions among diverse species. In a given ecosystem, all the living organisms partly or wholly interact with each other as well as with the surrounding air, water, soil and so on.

2. Distribution of Biodiversity

Distribution of biodiversity is not even on earth. In some regions it is consistently richer, while in some regions fewer species are found. Both plant and animal diversity depend much on a region's climatic condition, altitude, soil type and the existence of other species. Recently, large numbers of species are formally classified as endangered species. Besides, scientists estimate that many more endangered species actually are yet to be formally recognized. As of 2006, as per the criteria of IUCN Red List, with a total extinction of 16,119 species, about 40 percent of over forty thousand species assessed are listed as threatened species (Lovett, 2006). The most common practice of discerning global biodiversity is identification of biodiversity hotspots. By biodiversity hotspot, we understand a region possessing endemic species at a high level. Dr. Norman Myers identified these hotspots for the first time in two articles published in *The Environmentalist* (Myers, 1988 and 1990). It is unfortunate that such hotspots tend to occur in the vicinity of dense human habitation, which causes threats to many an endemic species. Owing to the pressures caused by ever increasing human population, there is dramatic increase of human activity in many such areas. These hotspots are mostly forests and are located in the tropics. For instance, the Atlantic forest in Brazil contains about 20,000 plant species and 1,350 vertebrates besides millions of insects, about 50 percent of which occur only in this region in the globe. The island of Madagascar, which includes the unique dry deciduous forests as well as lowland rainforests possess an incredibly high species-endemism-biodiversity ratio, since the island is isolated from mainland Africa for the last 65 million years, the species as well as ecosystems in the island evolved independently producing the species, which are conspicuously different from those found elsewhere in Africa. Many regions of the globe with high biodiversity coupled with high endemism arise from exceedingly specialized habitats requiring unusual mechanisms of adaptation. For instance, the peat bogs found in Northern Europe host a wide diversity of both flora and fauna, many of which are found nowhere else.

3. Evolutionary Perspective

Today's global biodiversity is the outcome of evolution that occurred for billions of years. Science is not certain about the actual origin of life, though limited evidence suggests that until about 600 million years back, all life forms consisted of bacteria and some similar unicellular organisms. From the apparent biodiversity revealed in the fossil record, we get the impression that in the past few million years, the greatest biodiversity occurred in the history of earth. However, all scientists do not agree to this, as there is great uncertainty regarding the degree of bias caused by the greater availability as well as preservation of fossil records from the recent geologic sections. According to some natural scientists, corrected for sampling artifacts, the preset-day biodiversity is not significantly different from that 300 million years ago. The present day global macroscopic species diversity is estimated to vary in the range of 2-100 million species, and the best estimate lies somewhere close to 10 million (J Alroy *et al.*, 2001).

However, biologists mostly agree that the period starting from the emergence of humans formed part of a new extinction on a massive scale, termed as, the Holocene extinction event. This was primarily caused by the impact that humans have on the environment. Presently, the estimated number of species that have become extinct owing to human action is still much smaller compared to that occurred during the major extinction events of the geological past. It is, however, argued that the current extinction rate is enough to create a massive extinction in under 100 years. While some others suggest that before the biodiversity loss matches the past losses during global extinction events to the tune of 20 percent or so, the present extinction rate could sustain for many thousands of years.

There are regular discoveries of new species, though not all of them are put into any taxonomic category yet.

4. Why Biodiversity Matters?

It can be sensed from the erstwhile discussion that there are manifold of benefits of global biodiversity in the sense that one diverse group always aids another. Throughout the history, the lack of diversity, popularly known as monoculture, was a factor contributing to several agricultural disasters. They include the collapse of European wine industry in the late nineteenth century, the Irish Potato Famine, and the epidemic caused by Southern Corn Leaf Blight of the United States in the second half of twentieth century (*cropdisease.cropsci.uiuc.edu:* official website of the University of Illinois). Higher biodiversity is also responsible for controlling the spread of diseases. For instance, the virus causing Lyme disease can be controlled in the presence of more different animal species surrounding its environment.

Biodiversity provides food for all living organisms including humans. We get a vast majority of our food from about 20 types of plants. Though we can use many kinds of animals as food, our consumption is mostly focused on a few species. If we can halt the current rate of extinction, there is an enormously vast untapped potential for widening the variety of food products that are suitable for our consumption. A considerable proportion of medicines are derived from biological sources, either directly or indirectly. In laboratory settings, such medicines cannot be synthesized in most cases at present. Besides, only an insignificant proportion of the entire plant diversity has been investigated thoroughly for potential drug sources. At that, many medicines as well as antibiotics are derived from different microorganisms.

A wide variety of industrial materials are directly derived from living organisms. These materials include inter alia building materials, adhesives, dyes, fibers, resins, rubber and oil. Potential for further quest into materials that can be utilized in a sustainable manner from a wider variety of organisms is also quite encouraging. In presence of a rich biodiversity, much technological advancement could be made possible, which would not be thought of otherwise. For certain plants having economic importance like food-crops, wild varieties of the existing domesticated crops can often be reintroduced so that some better varieties compared to the previously domesticated ones are formed. The economic impact is enormous, even for common crops like potato, which was bred initially through a single variety. Climate change can cause the wild varieties of potato suffer enormously.

Over and above, biodiversity provides various ecosystem services, which are not readily visible. Our atmospheric chemistry and water supply is regulated to a great extent by biodiversity. It is directly occupied in recycling nutrients as well as ensuring soil fertility. It is not easy for human beings to build ecosystems to support their needs. For instance, man-made construction cannot imitate insect pollination, an activity which singly represents enormous wealth in ecosystem services to mankind every year.

Furthermore, many of us derive value from earth's biodiversity through leisure activities like enjoying countryside walk, watching the picturesque surroundings or programs on natural history on television. Biodiversity also reportedly inspires artists like musicians, painters, sculptors and writers. Some cultural groups often view themselves as integral to the natural world and therefore show respect towards other living organisms.

5. Threats to Global Biodiversity

It is a well established fact that the global biodiversity is fraught with manifold threats posed in varied dimensions. The factors threatening biodiversity in estuaries or in the oceans are usually the same as those affecting biodiversity in terrestrial systems. Alien species introduction, physical change in habitat areas, overexploitation, and alteration in atmospheric composition all pose threats to biodiversity. The last century has increasingly witnessed erosion of biodiversity. Some studies show that every eighth species of plant is threatened with extinction. According to Pimm *et al.*, the annual biodiversity loss in a given year may go up to 1,40,000 species (Pimm *et al.*, 1995). This is quite indicative of unsustainable ecological practices, as only a small number gets added to the species diversity each year. Scientists mostly acknowledge that the rate at which species loss takes place is greater at present than at any other time in the history of mankind, so much so that the rate of extinction are hundreds of times higher when compared to background extinction rates (Thackery, 1990; Raymond and Ward, 2005).

Species extinctions occur mostly due to human activities, particularly destruction of habitats for plants and animals. The ever-increasing extinction rates are being triggered by continuous human consumption (Ehrlick and Ehrlick, 1981). While most of the currently endangered species are not food species, the biomass available in them can often be converted into human food after their habitat gets transformed into cropland, orchards, pasture and so on. A substantial portion of the global biomass is reportedly tied up in only a handful of species representing crops, livestock and humans. As the stability of an ecosystem decreases owing to the extinction of its species, we need to caution ourselves that our ecosystem is destined to collapse unless we succeed in preserving its complexity, particularly the bio-complexity. Environmental pollution, abrupt climatic change, overpopulation, deforestation and many other factors induced by human activity are responsible for biodiversity loss. Most of these factors stem from overpopulation and produce their cumulative impact upon global biodiversity.

Sometimes, biodiversity loss is characterized by conversion to trivial ecosystem standardization rather than ecosystem degradation. Lack of clarity regarding property rights or regulation concerning access to biotic resources can also lead to loss of biodiversity. In many parts of the globe, a rich species diversity of unique varieties exist only because of its separation caused by barriers like deserts, mountains, large rivers, seas and oceans from other species occupying separate land masses, especially the highly productive, "super-species". Natural processes would never be mighty enough to cross such barriers barring many millions of future years through continental drift. Still, mankind is vested with the power to bring forth many changes hitherto unclaimed in its evolutionary history, and in times to come, unlike the past centuries, which only involved major animal migrations.

Another potent threat to global biodiversity is the widespread introduction of numerous exotic species by humans. With the introduction and subsequent establishment of self-sustaining populations of exotic species into an ecosystem not evolved enough to cope with such species, survival of that ecosystem will be at immediate stake. The exotic species may turn out to be predators, parasites, or those depriving indigenous species of the essential nutrients, water and light. Such species often have features making them competitive against the existing ones owing to their evolutionary background. Consequently, if species hailing from different eco-regions are combined, there is a risk that relatively fewer aggressive species would dominate the global ecosystems.

6. Biodiversity Loss: Potential Threat to Ethnomedicine

Numerous wild species are overexploited all over the world, thanks to the demand fashioned by the traditional medicine. At this juncture, documentation of the traditional usage of flora and fauna in traditional medicine coupled with the cultural as well as ecological aspects associated with ethnomedicinal practices are of utmost importance (Alves and Rosa, 2005). It is quite obvious that the practice of traditional medicine is not free from the current environmental emergency facing the globe. Substantial changes in various types of vegetations have profound impact on both procurement and preparation, besides the cost of medicines extracted from plants. It is also argued that desecration of sacred spaces, spiritual spots, grooves etc., tend to hamper the dignity of those 'landscapes' and consequently encourage their abuse (Anyinam, 2005). For over a couple of decades now, degradation of Amazonian forests in Brazil has been responsible for diminishing the availability of medicinal plant species a great deal. The Amazonian forest degradation may signify the loss of prospective pharmaceutical drugs as well as the erosion of the region's sole health care option catering to the need of many poor (Shanley and Luz, 2003).

Triggered by the impact of rapid industrialization and urbanization, indigenous medical systems are increasingly being displaced by western medicine in many parts of the world, and eventually leaving many poor people with no health care. Knowledge of traditional medicine is disappearing at a rapid pace due to cultural change as well as dwindling access to natural medicinal sources. Globally, many villages no longer have the surrounding natural habitats previously serving as a medicine cupboards. At that, wealth of folk knowledge accumulating and being honed for centuries are disappearing alarmingly. Though in some cases such a loss may confer net health benefits, it is unlikely that modern society will ever realize the degree of loss being occurred to effective medicinal treatments

(Daily and Ehrlich, 1996). For instance, in Latin America, in spite of many government level biodiversity-preservation efforts for future generations, indigenous knowledge, particularly that derived from traditional medicine is rapidly disappearing (Calixto, 2005).

Local ecosystem transformation shaped through economic activities has allegedly been imposing severe bottlenecks on the availability as well as accessibility of specific kinds of species having medicinal value. Ecosystem degradation has been causing gradual disappearance of certain plant species thereby creating a situation, which will pose formidable challenge to the future ethno-medicinal practices. Barring a few exceptions, concoctions for all medicines have their base in plants, their organs or secreted products (Kerharo, 1975). Currently, species procurement required by indigenous medical practitioners need long distance travel, which affects operational costs of providing ethno-medicinal services as well as the forms of medicine prepared. For instance, various concoctions and tinctures are increasingly replacing freshly prepared herbal medicines for their durability (Anyinam, 1987).

As pointed out by Alves and Rosa (2007) though traditional medicine is important for public health in different regions of the globe, the ethnomedicinal practitioners are at a much higher risk of extinction than the forests and other biomes. Indigenous knowledge about many plants' medicinal use is disappearing more rapidly than the plants themselves. Tropical forest destruction has meant, in many a case, increasing disappearance of indigenous communities inhabiting those areas and accumulating a compendium of traditional knowledge about the curative value of various medicinal plants. Communities having extensive knowledge regarding local ecosystem may be involved for collaborative conservation as well as management. Such examples can be seen in Brazil (Castello, 2004; Gillingham,2001), Zimbabwe (Child, 1996), the Philippines (Pomeroy and Carlos, 1997), and in Pacific Islands (Michael and Lambeth, 2000). On the other hand, exclusion of indigenous communities from the process of consultation and decision-making can lead to the formulation of public policies having no historical information, and for obvious reasons without adequate social resonance.

7. Global Biodiversity Management: Measurement, Conservation and Critics

Within the broad embodiment of biodiversity, an array of objective measures has been created to measure biodiversity empirically. Each measure relates itself to a particular application of the data. A practical conservationist uses this measure to quantify a value which is shared broadly among the people who are affected locally. For others, biodiversity needs a more economically defensible understanding that enables ensuring of unrelenting possibilities for adaptation as well as future use by people, thereby assuring environmental sustainability. Consequently, biologists insist on measuring biodiversity in association with genetic diversity. Since it is quite difficult at times to identify the genes that could prove beneficial, the best conservation option is to ensure that more and more genes persist. On the other hand, the ecologists consider this latter approach as too restrictive, as it may inhibit ecological succession.

Usually, we plot biodiversity as taxonomic richness of an area, with reference to some chronological scale. According to Whinker (1972), the most common metrics that can be used for measuring species-level biodiversity, should be through focusing attention to the richness or evenness of species. Among all the available indices, species richness is recognized as the most primitive as well. Besides, there are three more indices used by ecologists *viz.*, alpha diversity that refers to diversity of a particular locality, community or ecosystem, beta diversity referring to diversity of species between ecosystems and gamma diversity, which measures the overall biodiversity for various ecosystems within a particular region.

The conservation of biodiversity has presently become a global concern owing mainly to its immense significance and inherent linkage with human existence. Though there are continuous debates on the extent as well as significance of current extinction, biodiversity is mostly considered essential, whereupon the need for conserving and managing it goes without much saying. There are mainly two types of prevailing conservation options, *viz.*, in-situ and ex-situ conservation. Though its implication is not always feasible, in-situ conservation is generally considered as an ideal conservation strategy. For instance, ex-situ conservation efforts are required at times in case of destruction of the habitats of endangered species. Besides, such efforts can act as backup solutions to in-situ initiatives.

To some conservationists, both types of conservation efforts are necessary for ensuring proper preservation. Setting up a protection area forms a common example of in-situ conservation effort. On the other hand, ex-situ conservation efforts include planting germplasts in seedbanks, growing unique species in nurseries and so on, which allow large scale preservation with minimal genetic erosion.

The conservation of biological as well as genetic diversity is one of the major goals of different levels of reserve systems. According to the 'International Union for the Conservation of Nature and Natural Resources' for adequate protection of biological as well as genetic diversity of any plant community we need a threshold of at least 12 percent. However, if we choose locations with no consideration to the intra-community variations, area-based thresholds may only be able to protect a small share of the total diversity. In this case, a coarse-filter approach may be applied for identifying gaps existing in a plant community's current reserve system by considering current climatic conditions and the scenario of global climate change as the bases (Coulston and Riitters, 2005).

Nationally, sometimes Biodiversity Action Plans are meant for stating the protocols required to individual species protection. Such plans also generally detail the extant data on individual species and its habitat. This is referred to as 'recovery plan' in some countries. There are several international instruments that accommodate biodiversity as a special component of discussion as well as action. The threat to global biodiversity is among the hot topics that are discussed in different international summits with the aim to see the formation of a global conservation initiative for helping maintain the biological wealth. For insuring biodiversity in marine as well as terrestrial systems, we need to recognize the fact that the entire natural resource base, people's economic development, our basic needs like food, medicine, clothing, the air we need for breathing, and to top it all our very existence on earth depends on the life surrounding us. Therefore, in preserving biodiversity, we are insuring nothing but our own continued survival and prosperity.

Some thinkers challenge the notion regarding untapped potential intended to reduce human dependence on a comparatively small number of domesticated species of both animals and plants. To them the rarity of species that could be domesticated and their scarce occurrence limits the number of locations where major civilizations could take place. Till recent past, despite extensive research on minor food sources, there is no instance of those becoming major food crops (Diamond, 1998). The province of biodiversity research is often fraught with natural human egocentric biases. It often faces criticisms for being overly guided by the founders' interests giving a narrow focus instead of extending to potentially useful areas (Irish and Norse, 1996; France and Rigg, 1998). Mobilization of public opinion as well as national legislation is seemingly easier for terrestrial biodiversity as compared to the marine ones. The terrestrial realm has a considerably higher visibility as it falls within the distinct boundary of a particular country. Marine conservation, on the other hand, involves adoption of fresh international mechanisms for both protection and addressing methodological issues in the discourse relating to the classification of marine ecosystem and gathering of information on some of the most difficult species in the globe to access as well as to monitor the same.

8. Judicial Perspective

Over and above its evaluation from an evolutionary point of view, biodiversity is also being taken into account of late in political as well as judicial decisions. The ecosystem-law relationship can be traced back to history and this relationship has important consequences for global biodiversity. It relates to public as well as private property rights. It can also define protection for endangered ecosystems, in addition to some relevant rights and duties such as fishing rights, hunting rights and so on. Law concerning species is a relatively recent issue, which defines species requiring immediate protection because of threats posed by extinction. The Endangered Species Act of the United States is one such example. Application of these laws is, however, always subject to scrutiny.

Laws concerning gene pools have also been instrumental with the progress that the genetic field has experienced in the past few decades. This progress has reportedly led to tightening of relevant laws in this field. As the emerging technologies of genetic analysis as well as genetic engineering are making strides, gene patenting and process patenting have become very familiar issues. Among the hottest debates today, one seeks to define whether a particular resource is the organism, or its gene, or its DNA. The UNESCO convention of 1972 established that biological resources like plants were mankind's common heritage. Such rules might have inspired the creation of grand public genetic resource banks, situated outside the source-countries.

More recent global agreements such as the Convention on Biological Diversity (CBD) give sovereign rights over biological resources at the national level. The concept of static biodiversity conservation is gradually being replaced by that of dynamic conservation, with the idea of resource and innovation. Different recent agreements commit nations to biodiversity conservation, resource development for ensuring sustainability and sharing of benefits resulting from the use of those resources. It may be expected that under new rules, bio-prospecting or collecting natural products should be allowed by the countries with rich biodiversity through benefit-sharing.

Access and Benefit Sharing Agreements (ABAs) form the hallmark of sovereignty principles. The CBD spirit hinges upon a prior informed consent involving the collector and the source country, to establish the type of resource as well as purpose of using the same, in addition to effectuate a settlement on a fair benefit-sharing agreement. Any sort of disregard to such principles can cause bio-prospecting turn into bio-piracy. Uniform approval for using global biodiversity in the form of a legal standard is yet to be achieved, however. Some legal commentators also opine that one should not use biodiversity as a legal standard, as the different layers of innate scientific uncertainty pertaining to the biodiversity concept would cause administrative waste and trigger litigation impeding preservation goals (Bosselman, 2004).

9. CBD and TRIPs: Compatibility Versus Contradiction

The Convention on Biological Diversity (CBD), which was adopted in 1992, aims at securing the conservation as well as sustainable use of earth's biological diversity. Following this, the "Agreement on Trade Related Aspects of Intellectual Property Rights" (TRIPs) was concluded in 1992 in the agreement package of the World Trade Organisation (WTO). This Agreement sets minimum standards concerning Intellectual Property Rights (IPRs) including patents in the 134 signatory countries. The intricate legal as well as socio-political links between

IPRs and the conservation of biological diversity including genetic resources are quite explicit in the biotechnology sector, as genetic resources provide knowledge as well as the raw material required by the biotechnology industry. In the event of knowledge and information in a regulated market turning into saleable products, individual plants as well as animals may likewise get transformed from public goods to private ones. Thus, striking a balance between private and public interests become an international concern after the TRIPs agreement. There has been considerable debate on various fronts like environmental, ethical, and socioeconomic impacts of IPRs, particularly patents. The real knowledge enhancement regarding CBD-TRIPs relationship must be informed by solid evidence obtained from different countries through case studies, impact assessments and the like. The relationship is multifaceted as well as complex. The objectives of the CBD are: "the conservation of biological diversity, the sustainable use of its components and the fair and equitable sharing of the benefits arising out of the utilization of genetic resources, including by appropriate access to genetic resources and by appropriate transfer of relevant technologies, taking into account all rights over those resources and to technologies, and by appropriate funding" (Source: *www.cbd.int*). On the other hand, the TRIPs Agreement aims at a multilateral framework to promote effective as well as adequate protection of IPRs for minimizing distortions and impediments to global trade and ensuring that measures and procedures for enforcing IPRs themselves do not act as barriers to trade. According to Article 7 of the Agreement, "protection and enforcement of intellectual property rights should contribute to the promotion of technological innovation and the transfer and dissemination of technology, to the mutual advantage of producers and users of technological knowledge and in a manner conducive to social and economic welfare, and to a balance of rights and obligations" (*www.wto.org*).

Though the Agreement on both CBD and TRIPs, have so far come more or less parallelly, there are reasons which merit a re-examination of their compatibility. For instance, both are the outcomes of the multilateral system, whereupon any point of inconsistency arising in the two must be addressed in a way that enables the signatory countries to meet the compliance requirements for both the Agreements.

10. Conclusion

The conservation of biodiversity has presently become a global concern owing mainly to its immense significance inherent linkage with human existence. Though there are continuous debates on extent as well as significance of current extinction, biodiversity is mostly considered essential, whereupon the need for conserving and managing it goes without much saying. Biodiversity is also being taken into account of late in political as well as judicial decisions. Some global agreements give sovereign rights over biological resources at the national level. Numerous wild species are overexploited all over the world, mainly because of the demand fashioned by the traditional medicine, which necessitates proper documentation of the indigenous usage of flora and fauna in traditional medicine coupled with the cultural as well as ecological aspects associated with ethno-medicinal practices.

(Azmal Hussain, Faculty Associate, The Icfai Business School Research Centre, Kolkata. He can be reached at azmalhussain@icfai.org).

Literature Cited

Alves R R N and I L Rosa 2005, Why Study the Use of Animal Products in Traditional Medicines? *Journal of Ethnobiology and Ethnomedicine*. 1:1-5.

Alves R N R and I M L Rosa 2007, Biodiversity, Traditional Medicine and Public Health: Where do they Meet?. *Journal of Ethnobiology and Ethnomedicine*. 3: 14

Anyinam C 1995, Ecology and Ethnomedicine: Exploring Links Between Current Environmental Crisis and Indigenous Medical Practices. Social Science & Medicine 40(3):321-329.

Anyinam C 1995, Ecology and Ethnomedicine: Exploring Links Between Current Environmental Crisis and Indigenous Medical Practices. Social Science and Medicine 40(3):321-329.

Anyinam C A 1987, Persistence with Change: A Rural-Urban Study of Ethno-Medical Practices in Contemporary Ghana. PhD thesis. Queen's University, Kingston, Ontario (cited from Alves and Rosa, 2007)

Bosselman F 2004, A Dozen Biodiversity Puzzles. N.Y.U. *Environmental Law Journal* 12: 364.

Calixto J B 2005, Twenty-five years of Research on Medicinal Plants in Latin America. *Journal of Ethnopharmacology.* 100:131-134

Castello L 2004, A Method to Count Pirarucu Arapaima Gigas: Fishers, Assessment and Management. *North American Journal of Fisheries Management.* 24:379-389

Child B: The Practice and Principles of Community-based Wildlife Management in Zimbabwe: the CAMPFIRE Programme. Biodiversity and Conservation. 5(3):369-398.

Coulston J W and K H Riitters 2005, Preserving Biodiversity under Current and Future Climates: A Case Study. Global Ecology and Biogeography, 14, 3138 (accessed on August 6, 2007 from *http://www.treesearch.fs.fed.us/pubs/7834).*

Daily G C and P R Ehrlich 1996, Global Change and Human Susceptibility to Diseases. Annual Review of Energy and Environment. 21:125-44

Diamond J 1998, Guns, Germs and Steel. Vintage (cited from *http://en.wikipedia.org/wiki/Biodiversity)*

Ehrlich P and A Ehrlich 1981, Extinction. Random House. New York.

France R and C Rigg 1998, Examination of the 'Founder Effect' in Biodiversity Research: patterns and imbalances in the published literature. Diversity and Distributions. 4: 77-86

Gaston K J and J I Spicer 2004, Biodiversity: An introduction. Blackwell Publishing.

Gillingham S, 2001. Social Organization and Participatory Resource Management in Brazilian Ribeirinho Communities: A Case Study of the Mamirauá Sustainable Development Reserve, Amazonas. Society and Natural Resources 14(9):803-814.

Irish K E and E A Norse 1996, Scant Emphasis on Marine Biodiversity Conservation Biology. 10: 680.

J Alroy C R, *et al.*, 2001. Effect of Sampling Standardization on Estimates of Phanerozonic Marine Diversification. Proceedings of the National Academy of Science, USA 98: 6261-6266 (cited from *http://en.wikipedia.org/wiki/Biodiversity).*

Kerharo J 1975, Traditional Pharmacopoieas and Environment. African Environment 1:30.

Lovett R A 2006, Endangered Species List Expands to 16,000. National Geographic News. (accessed on August 22, 2007 from *http://news.nationalgeographic.com/news/2006/05/0502_060502_endangered.html).*

Michael K Lm, Lambeth L 2000, Fisheries Management by Communities: A manual on, promoting the management of subsistence fisheries by Pacific Island communities. Secretariat of the Pacific Community (cited from Alves and Rosa, 2007).

Myers N 1988, Threatened biotas: 'hot spots' in tropical forests. The Environmentalist 8: 187-208.

Myers N 1990, The Biodiversity Challenge: Expanded Hot-spots Analysis. The Environmentalist, 10: 243-256.

Pimm S L, G J Russell, J L Gittleman and T M Brooks 1995, The Future of Biodiversity. Science 269: 347-350.

Pomeroy R S and M B Carlos 1997, Community-based Coastal Resource Management in the Philippines: A Review and Evaluation of Programs and Projects, 1984–1994. Marine Policy. 21(5):445-464

Raymond H and P Ward 2005, Hypoxia, Global Warming, and Terrestrial Late Permian Extinctions. Science 15: 389–401.

Shanley P and L Luz 2003, The Impacts of Forest Degradation on Medicinal Plant use and Implications for Health care in Eastern Amazonia. BioScience 2003. 53(6):573-584.

Thackery J 1990, Rates of Extinction in Marine Invertebrates: Further Comparison between Background and Mass Extinctions". Paleobiology 16: 22-24.

United Nations Educational, Scientific and Cultural Organisation 1972, Convention Concerning the Protection of the World Cultural and Natural Heritage. Adopted by the General Conference at its seventeenth session Paris, 16 november 1972. (accessed on August 24, 2007 from *http://whc.unesco.org/archive/convention-en.pdf*).

Whit taker RH 1972, Evolution and Measurement of Species Diversity. Taxon. 21: 213-251.

Wilson E O and Frances M Peter (eds), 1988. Biodiversity. National Academy Press (cited from *http://en.wikipedia.org/wiki/Biodiversity*).

Websites

http://www.cbd.int/convention/convention.shtml

http://www.cbd.int/doc/legal/cbd-un-en.pdf

http://cropdisease.cropsci.uiuc.edu/corn/southerncornleafblight.html

http://www.fws.gov/Endangered/esa.html (official website of US Fish and Wildlife Service)

http://cropdisease.cropsci.uiuc.edu/corn/southerncornleafblight.html

http://www.unctad.org/trade_env/docs/cbd-trip.pdf

http://www.cbd.int/convention/articles.shtml?a=cbd-01

http://www.wto.org/english/tratop_e/trips_e/trips_e.htm

http://www.sms.si.edu/IRLSpec/Whats_biodiv.htm

2

Policy Brief
Preserving Biodiversity and Promoting Biosafety

This article is based on the opinion that governments should aim to maintain biological diversity even as encouraging economic growth. Once governments decide to act to protect biodiversity, they need to choose policies that reflect the relative value people place on it. Incentive measures are the basis of a market approach to biodiversity management that help reconcile differences between the value of biodiversity-related resources to individuals and the value of biodiversity to society as a whole. Governments have two important roles in supporting markets for biodiversity-related resources. These are: establishing the right framework conditions for private and public operators to supply biodiversity-related resources efficiently to users; and applying the right policy instruments to ensure that public biodiversity-related goods and services are provided in the most efficient and effective manner.

Introduction

Biodiversity – the variety of life and of habitats on Earth – is vital to human welfare. The loss or degradation of biodiversity can have important economic, environmental, and social consequences. Altering a watershed (the area draining into a common waterway), for example, not only leads to the potential loss of an ecosystem – through loss of habitat – but may also create economic costs for water filtration in cities using its water.

And it is not enough simply to preserve those biological resources that are known to be useful to humans now – we may also be losing potentially beneficial compounds and materials that are, as yet, undiscovered (e.g., genetic resources for use in pharmacological or agricultural applications).

Biodiversity loss also has social consequences, in its impact on people's livelihoods and lifestyles – this is the unmeasured cost of losing cultural traditions.

The loss of key elements of an ecosystem can alter the balance between its components and lead to long-term or permanent changes. Biodiversity loss can also affect human health, as our health is largely dependant on the quality of the ecosystem in which we live – loss of plant or animal species can affect the quality of water or soil, for instance.

The main pressures on biodiversity result from land use changes (usually associated with increasing populations); unsustainable use and exploitation of natural resources (especially fisheries, agriculture, and forestry); global climate change; and industrial pollution. At the same time, biotechnology is introducing new organisms and their effect on existing organisms and habitats also needs to be considered.

In some instances, these pressures can actually be positive for biodiversity. Agricultural activity sometimes improves the habitat and even helps increase the variety of species; the Mediterranean basin is considered a biodiversity "hot spot" in part because of its human-induced agricultural biodiversity.

However, the available evidence suggests that, in most regions of the world, the effects of economic activity are negative for biodiversity. This Policy Brief looks at what governments can do to place a value on and help preserve biodiversity and ensure biosafety.

What is Biodiversity Worth?

Few would dispute that governments should aim to maintain biological diversity as well as encourage economic growth. But how much biodiversity protection is needed and what is the most efficient way to deliver it? These are fundamentally economic questions, which is why the OECD has been examining the links between economic activity and biodiversity management for the past decade.

Some would argue that, as incomes rise over time, preferences for improved environmental quality lead automatically to better conservation and better use of biodiversity-related resources. Proponents of this argument note the improvement in air and water quality that has accompanied the advanced stages of industrialisation in most countries. This observation, however, seems limited to those environmental amenities where collective action is easily accomplished, such as air quality in large cities. Biodiversity in general is more dispersed and less easily targeted for collective policy. There is also less room for error – if air becomes more polluted, steps can be taken to reduce the pollution, but if an ecosystem is compromised by human encroachment, its loss may be permanent. Given that our knowledge of the complexity and inter-linkages of ecosystems are only cursory, loss of biodiversity may be more important than any immediate gains.

Before policy-makers take decisions about whether to risk destroying a habitat for economic gain (such as a housing development), they need to be able to set a relative value for those gains and for the biodiversity at risk. But most environmental amenities or biodiversity – landscapes, species of insect, forest habitats – are not bought or sold and therefore a price cannot be set on their use without some form of policy intervention. Without collective action, over-exploitation of such common assets is the likely result. In the case of biodiversity, at the extreme this leads to such destructive practices as "slash and burn" farming on fragile soils. Overall, there is a good case to be made for vigorous policy intervention to help preserve both the extent and the quality of biodiversity.

What Policies for Biodiversity?

Once governments decide to act to protect biodiversity, they need to choose policies that reflect the relative value people place on it. Placing a market value on biodiversity may sound difficult, but the strength of people's desire to maintain biodiversity can be tested at many levels. For example, evidence can be gleaned from people's willingness to visit natural areas and the pleasure they derive from nature hikes and other non-destructive outdoor activities. The rapid growth of the global eco-tourism industry also demonstrates a strong willingness by people to pay substantial amounts of money to see natural phenomena that are not available locally. Many studies have been conducted to quantify the value people place on biodiversity-related goods and services relative to other economic outputs.

Governments can use this information to craft public policy that will encourage the use and conservation of biodiversity in a way that reflects its relative value. OECD work on the economic aspects of biodiversity explores the types of policies and instruments that will lead to uses of biodiversity-related resources that best reflect relative preferences. Part of this work seeks to ensure that the biodiversity-related impacts of economic activity are an explicit part of the day-to-day decisions that consumers and producers face. For example, imposing a higher cost on products using virgin resources than those using recycled resources could be used to reflect impacts on biodiversity. When biodiversity policy accounts for both private and public values of biodiversity, as well as for the consequences on all affected individuals (including future generations), the use of biodiversity resources will be consistent with achieving the greatest net benefit to society over the long-term.

The choice of policy instruments is complex and depends on specific institutional, economic and social needs. Policy options should be systematically analysed to minimise the costs of public administration, monitoring and enforcement, as well as the private costs of implementation. Since market-based instruments that change the prices of biodiversity-impacting products (e.g., wood products) are often cost-effective – and generally under-utilised – they should be promoted. However, it will still often be necessary to use non-market-based instruments (e.g., regulatory initiatives such as banning trade in endangered species) in the policy mix.

There is also a need to work at the international level to implement biodiversity management policies, for example in development co-operation or in policies to protect migratory species and aquatic resources. Moreover, biodiversity-related resources have non-use values to people everywhere which need to be reflected in local use decisions – without compromising local economic development.

How to Encourage Good Biodiversity Use?

Incentive measures are the basis of a market approach to biodiversity management. Such measures help reconcile differences between the value of biodiversity-related resources to individuals and the value of biodiversity to society as a whole. They increase the cost of activities that damage ecosystems important for biodiversity and reward biodiversity conservation and enhancement/restoration. Farmers who receive a government payment for maintaining biological diversity on their land, for example, will be more willing to use farm practices that sustain biodiversity values.

The idea of using incentives to achieve particular biodiversity objectives often requires a method for gauging the value of biodiversity. Unfortunately, reaching consensus on such method(s) is not always easy. Economic valuation can sometimes offer a solution if monetary measures of the impacts involved can be obtained or implied – such as the cost of travel, and related expenses, for visiting natural areas. However, it is more difficult to quantify the more aesthetic or cultural values involved, for example, in non-use of biodiversity, such as declaring a mountainside an ecological conservation area. For those non-use values, techniques to "reveal" biodiversity's value to people are needed.

Although some debate continues about how far economic valuation techniques can be used to value non-marketed environmental resources such as landscapes, the acceptability and use of these techniques continue to grow. This is mainly due to theoretical advances in the methodologies underlying these techniques.

How to Create Markets for Biodiversity?

Markets are created by removing barriers to trade, including the establishment and assignment of well-defined and stable property and/or user rights. Market creation is based on the premise that holders of these rights will maximise the value of their resources over time, thereby optimising biodiversity use, conservation, and restoration. Market creation therefore involves a broader approach than the simple use of market incentives.

Governments have two important roles to play in supporting markets for biodiversity-related resources. First, they need to establish the right framework conditions for private and public operators to supply biodiversity-related resources efficiently to users. They also need to apply the right policy instruments to ensure that public biodiversity-related goods and services are provided in the most efficient and effective manner.

Three issues need special attention. First, the absence of appropriate information can inhibit the development and implementation of market approaches to biodiversity conservation, use, and restoration. Information can be provided through such mechanisms as labelling, certification and technical capacity-building. Scientific knowledge is also important, so governments need to develop policies that establish the right conditions for new knowledge to emerge in relation to biodiversity conservation. Indicators to monitor biodiversity change will also be important, along with the active and early engagement of stakeholders in developing and implementing biodiversity management policies. Local community networks that identify and support local biodiversity objectives can make important contributions in this regard. Also, markets need to be periodically monitored to ensure they actually result in net benefits for society as a whole.

A number of specific markets have already developed around biodiversity-related activities. Examples include: organic agriculture; sustainable forestry; non-timber forest products; genetic resources; and eco-tourism. Two highly successful examples where the instruments themselves created the market are trading in access to fishing rights and transferable development rights to land. The emergence of private parks in many regions of the world also demonstrates that there is scope for capturing public values in private markets. For those parks, the private value of their uniqueness is high enough to support public biodiversity objectives. However, since the public value of the parks will typically be greater than the private values, economic incentives that capture some of these additional public values would improve the efficiency and effectiveness of biodiversity management.

Why is Biosafety Important?

Products of modern biotechnology might also affect biodiversity, and thus raise environmental safety issues. Genetically engineered crops (also known as transgenic crops) are no longer an idea of the future, but have well and truly arrived. Crops such as maize, soybean, rapeseed and cotton are being approved for commercial use in an increasing number of countries. The OECD has been active on such issues since the mid-1980s.

From 1996 to 2004, there was more than a 47-fold increase in the area grown with transgenic crops worldwide, reaching 81.0 million hectares. In 2004, there were 14 countries that grew 50,000 hectares or more: the US grew 59% of the world total, followed by Argentina (20%), Canada (6%), Brazil (6%), China (5%), Paraguay (2%), India (1%), and South Africa (1%). In addition, Uruguay, Australia, Romania, Mexico, Spain and the Philippines each had smaller scale cultivation of less than 1% of the total. Most of this cultivation was devoted to soybean (60%), maize (23%), cotton (11%), and rapeseed (6%). So far, most commercialisation has focused on these crops, and the genetic engineering has involved two traits: insect resistance and herbicide tolerance (Clive James, International Service for the Acquisition of Agri biotech (ISAAA), 2004).

The trend for an increase in the transgenic crop area seems set to continue, given the large range of genetically engineered crops in research and development

In all OECD countries, a notification or registration has to be made to obtain approval for the commercial use of a genetically engineered crop – whether for planting and growing or for use in human or animal foods. Such approval is always based on a safety assessment by national authorities, which in turn is based on scientific information regarding the crop, its specific trait, and the receiving environment. It is important for governments to ensure that good quality safety information is publicly available and, where possible, to adopt international approaches to risk and safety assessment that ensure the efficiency of the risk assessment process.

OECD Work Supports International Commitments on Biodiversity and Biosafety

OECD work on biodiversity and biotechnology helps countries to work together to develop effective and economically efficient measures to comply with their international commitments in these areas.

Many of the priorities of OECD countries related to the economics of biodiversity and biosafety are expressed within the Convention on Biological Diversity (CBD) and its Biosafety Protocol. For this reason, much of the work of the OECD in both areas is channelled toward the CBD Secretariat, and a strong co-operative relationship between the two Organisations has evolved.

The result has been that many OECD outputs, such as those regarding incentive measures, have also found their way into decisions generated by the Conference of Parties to the CBD itself.

How to Ensure Biosafety?

One of the primary goals of the OECD's work on biosafety in the past decade has been to promote harmonisation among member countries of notifications and registrations of biotechnology products. Such harmonisation aims to ensure that the information used in risk/safety assessments, as well as the methods used to collect such information, are as similar as possible.

This can lead to countries recognising or even accepting information from one another's safety assessments, and it generates significant benefits. It increases mutual understanding among member countries of each other's risk assessments; it avoids duplication of effort; it saves on scarce resources; and it increases the efficiency of the risk/safety assessment process. This in turn improves safety, while reducing unnecessary barriers to trade.

For harmonisation to be possible among member countries, it is important that they have similar approaches to risk/safety assessment. Earlier OECD work on risk assessment demonstrated that such assessment should be based on the characteristics of the organism, the introduced trait, the environment into which the organism is introduced, the interaction between these, and the intended application. This work has formed the basis for environmental risk/safety assessment that is now globally accepted.

The similarity of approach was reinforced by the fact that most genetically engineered organisms are developed from organisms such as crop plants whose biology is well understood. This allows the risk assessor to draw on previous knowledge and experience with the introduction of plants and micro-organisms into the enviro2nment. The process takes account of a wide range of attributes including, for example, knowledge and experience with the plant, including its flowering/reproductive characteristics, ecological requirements, and past breeding experiences.

It is not just that national authorities use the same concepts and principles in risk assessment. OECD countries also share a remarkably high degree of similarity in the questions and issues addressed in risk/safety assessments as outlined in national laws, regulations, and guidance documents. Of course, national authorities also request some regulatory information, which is specific to the local environment. But nevertheless, much of the information used in risk/safety assessment that relates to the biology of crop plants and micro-organisms is similar or virtually the same in all assessments involving the same organism.

So a major focus of the OECD's work on harmonisation is to compile the biological information common to the risk/safety assessment of a number of transgenic products, focusing on two specific categories: the biology of the host species or crop; and traits used in genetic modifications. The aim is to encourage information-sharing and prevent duplication of effort among countries by avoiding the need to address the same common issues in each application involving the same organism or trait.

The resulting Consensus Documents on biology and traits are not intended to be a substitute for a risk/safety assessment, because they address only the generic part of the information that member countries believe is relevant to risk/safety assessment. They are intended to be a "snapshot" of current information, for use during the regulatory assessment of products of biotechnology. Nevertheless, they make an important contribution to environmental risk/safety assessment and help prevent duplication of effort among countries.

An additional challenge is to transfer the knowledge obtained by countries with experience in risk/safety assessment to all those who need it, including non-OECD countries. This enables those involved in risk assessment of transgenic products to easily obtain information developed from previous experiences. The OECD has established information exchange mechanisms and databases to support this process. One such mechanism, BioTrack Online, has been one of the best sources of information on regulatory developments in OECD countries, as well as field trials and approvals of commercial products. The OECD's Product Database *(www.oecd.org/biotrack/productdatabase)* contains information on the safety of genetically engineered crops in commercial use, as well as links to the Web sites of national authorities.

BioTrack Online is publicly available on the Internet, and in recent years has been linked with the Biosafety Clearing House (BCH), which is part of the Cartagena Biosafety Protocol to the UN Convention on Biological Diversity. The

Cartagena Protocol is intended to lay the foundation for a global system for assessing and managing the impact of living modified organisms on biodiversity.

How to Identify Genetically Engineered Crops?

Confusion can arise when national authorities share information on the same genetically engineered crop if different names or descriptions are used for the same type of maize or cotton. To avoid this problem, the OECD has developed a system of "unique identifiers" for transgenic plants. A unique nine-digit letter and number code is given to each new transgenic plant that is approved for commercial use and becomes its "name" worldwide. So, for instance, a maize developed by Monsanto to be resistant to insect pests has a unique identifier of MON-ØØ810-6, while DD-Ø1951A-7 denotes a cotton developed by DuPont.

The guidance provides for the developers of a new transgenic product to generate the identifier. Once approved, national authorities can forward the unique identifier for inclusion in the OECD's database.

OECD countries are already using the system. The EU recently adopted it as its system for generating unique identifiers and it has been recognised as a mechanism for unique identification to be used within the context of the Cartagena Protocol. The OECD has been forwarding unique identifiers to the Biosafety Clearing House *(http://bch.biodiv.org/)*, which is a key element of the Protocol's activities.

The system works well for genetically engineered crops and the OECD is now considering how the identifier tool can be extended beyond crops to micro-organisms and animals. Thanks to unique identifiers, all stakeholders, including the public, will be able to access solid and reliable information when making their judgements about safety.

For Further Reading

OECD (2005), An Introduction to the Biosafety Consensus Documents of OECD's Working Group on Harmonisation of Regulatory Oversight in Biotechnology, OECD: Paris.

James C (2004), "Global Status of Commercialized Biotech/GM Crops: 2004", ISAAA *(www.isaaa.org)*

OECD (2004), Recommendation of the Council on the Use of Economic Instruments in Promoting the Conservation and Sustainable Use of Biodiversity, OECD: Paris.

OECD (2004), Handbook of Market Creation for Biodiversity: Issues in Implementation, OECD: Paris.

OECD (2003), Harnessing Markets for Biodiversity: Towards Conservation and Sustainable Use.

OECD (2003), Perverse Incentives in Biodiversity Loss. ENV/EPOC/GSP/BIO(2003)2/FINAL.

OECD (2002), Handbook of Biodiversity Valuation: A Guide for Policy-Makers, OECD: Paris.

OECD (2002), OECD Guidance for the Designation of a Unique Identifier for Transgenic Plants, OECD: Paris.

OECD (2001), Valuation of Biodiversity Benefits: Selected Studies, OECD: Paris.

OECD (2001), Environmental Outlook and Strategy: Biodiversity. Document ENV/EPOC/GEEI/BIO(2001)2/FINAL.

Report of OECD's Working Group for Harmonisation of Regulatory Oversight in Biotechnology to the G8 Heads of State and Government (OECD. 2000) (C(2000) 86/ADD2).

OECD (1999), Handbook of Incentive Measures for Biodiversity: Design and Implementation, OECD: Paris.

OECD's Biosafety Consensus Documents are available at: *www.oecd.org/biotrack/*

World's Biodiversity Hotspots

Conservation International defines biodiversity hotspots as "earth's biologically richest places with high number of species". The organization so far has identified the following biodiversity hotspots around the globe:

1. Atlantic Forest
2. California Floristic Province
3. Cape Floristic Region
4. Caribbean Islands
5. Caucasus
6. Cerrado
7. Chilean Winter Rainfall-Valdivian Forest
8. Coastal Forests of Eastern Africa
9. East Melanesian Islands
10. Eastern Afromontane
11. Guinean Forests of West Africa
12. Himalaya
13. Horn of Africa
14. Indo-Burma
15. Irano-Anatolian
16. Japan
17. Madagascar and Indian Ocean Islands
18. Madrean Pine-Oak Woodlands
19. Maputaland-Pondoland Albany
20. Mediterranean basin
21. Meso-America
22. Mountains of Central Asia
23. Mountains of Southwest China
24. New Caledoria
25. New Zealand
26. Philippines
27. Polynesia-Micronesia
28. Southwest Australia
29. Succulent Karoo
30. Sundaland
31. Tropical Andes
32. Tumbes-Choco-Magdalena
33. Wallacea
34. Western Ghats and Sri Lanka

Source: Biodiversity Hotspots. Conservation International. CI Facts [accessed from http://web.conservation.org/ImageCache/news/content/press_5freleases/2005/february/hotspots2_5fkit/hotspots2map_2epdf/v1/hotspots2map.pdf

3

Biodiversity Loss Threatens Human Well-Being

Sandra Díaz, Joseph Fargione, F Stuart Chapin III and David Tilman

In this article, the authors provide a synthesis of the most crucial messages emerging from the latest scientific literature and international assessments, on the role of biodiversity in ecosystem services and human well-being. Many activities indispensable for human subsistence lead to biodiversity loss, and this trend is likely to continue in the future. Most of the concrete actions to slow down biodiversity loss fall under the domain of policy-making by governments and the civil society. Though it has been less recognized, biodiversity also influences human well-being, including the access to water and basic materials for a satisfactory life, in the face of growing environmental changes. The environmental changes have large impacts on ecosystem processes, and thus human well-being.

The diversity of life on earth is dramatically affected by human alterations of ecosystems (1) Compelling evidence now shows that the reverse is also true: biodiversity in the broad sense affects the properties of ecosystems and, therefore, the benefits that humans obtain from them. In this article, we provide a synthesis

Source: PLOS Biology, August 2006.

of the most crucial messages emerging from the latest scientific literature and international assessments of the role of biodiversity in ecosystem services and human well-being.

Human societies have been built on biodiversity. Many activities indispensable for human subsistence lead to biodiversity loss, and this trend is likely to continue in the future. We clearly benefit from the diversity of organisms that we have learned to use for medicines, food, fibers, and other renewable resources. In addition, biodiversity has always been an integral part of the human experience, and there are many moral reasons to preserve it for its own sake. What has been less recognized is that biodiversity also influences human well-being, including the access to water and basic materials for a satisfactory life, and security in the face of environmental change, through its effects on the ecosystem processes that lie at the core of the systems (Figure 1).

Three recent publications from the Millennium Ecosystem Assessment (2-4), an initiative involving more than 1,500 scientists from all over the world (5), provides an updated picture of the fundamental messages and key challenges regarding biodiversity at the global scale. Chief among them are: (a) human-induced changes in land cover at the global scale lead to clear losers and winners among species in biotic communities; (b) these changes have large impacts on ecosystem processes and, thus, human well-being; and (c) such consequences will be felt disproportionately by the poor, who are most vulnerable to the loss of ecosystem services.

What We Do Know: Functional Traits Matter Most

Biodiversity in the broad sense is the number, abundance, composition, spatial distribution, and interactions of genotypes, populations, species, functional types and traits, and landscape units in a given system (Figure 2). Biodiversity influences ecosystem services, that is, the benefits provided by ecosystems to humans, that contribute to making human life both possible and worth living (4) (Box 1). As well as the direct provision of numerous organisms that are important for human material and cultural life (Figure 1, path 1), biodiversity has well-established or putative effects on a number of ecosystem services mediated by ecosystem processes (Figure 1, path 2). Examples of these services are pollination and seed dispersal of useful plants, regulation of climatic conditions suitable to humans and the animals and plants they consider important, the control of agricultural pests and diseases, and the regulation of human health. Also, by affecting ecosystem processes such as biomass production most vital by plants, nutrient and water cycling, and soil formation and retention, biodiversity indirectly supports the production of food, fiber, potable water, shelter, and medicines. The links between biodiversity and ecosystem services have been gaining increasing attention in the scientific literature of the past few years (2-4, 6). However, not until now has there been an effort to summarize those components of biodiversity that do, or should, matter the most for the provision of these services, and the underlying mechanisms explaining those links (Table 1; see also (3)).

From Ecosystem Processes to Human Well-Being

Ecosystem processes are intrinsic processes and fluxes whereby an ecosystem maintains its integrity (such as primary productivity, trophic transfer from plants to animals, decomposition and nutrient cycling, evapotranspiration, etc.). They exist independently from human valuation, and their magnitude and rate can be established regardless of the cultural, economic, and social values and interests of different human groups (Figure 1, Ecosystem Processes box).

Ecosystem services are the benefits provided by ecosystems that contribute to making human life both possible and worth living. Ecosystem services are context-dependent; that is, the same ecosystem process can produce an ecosystem service that is highly valued by one society or stakeholder group but not highly valued by other societies or groups. Some ecosystem services involve the direct provision of material and non-material goods and are associated directly with the presence of particular species of plants and animals—for example, food, timber, medicines, and ritual materials (Figure 1, path 1 and bottom sub-box of Ecosystem Services box). Other ecosystem services arise, either directly or indirectly, from the continued functioning of ecosystem processes. For example, the service of formation, retention, and sustained fertility of soils necessary for the production of plants and animals considered important by different human societies depends on the ecosystem processes of decomposition, nutrient cycling by soil microbiota, and the retention of water and soil particles by a well-developed root network (Figure 1, path 2 and top sub-box (#) of Ecosystem Services box). Some authors (e.g., [30]) have advocated a stricter definition of ecosystem services as components of nature that are directly enjoyed, consumed, or used in order to maintain or enhance human well-being. Although such an approach can be useful when it comes to ecosystem service accounting, our emphasis here is conceptual, and therefore we prefer to use the broader, widely accepted definitions and classification adopted by the Millennium Ecosystem Assessment [4]. This is because some ecosystem services (e.g., food provision) can be quantified in units that are easily comprehensible by policy makers and the general public. Others—for example, the services that regulate and support the production of tradable goods—are more difficult to quantify. If a criterion based on economic accounting is applied too strictly, there is a risk that

Contd...

Contd...

ecosystem service assessment could be biased toward services that are easily quantifiable, but not necessarily the most critical ones [29].

Human well-being is a human experience that includes the basic materials for a good life, freedom of choice and action, health, good social relationships, a sense of cultural identity, and a sense of security. The sense of well-being is strongly dependent on the specific cultural, geographical, and historical context in which different human societies develop, and is determined by cultural-socioeconomic processes as well as by the provision of ecosystem services. However, the well-being of the vast majority of human societies is based more or less directly on the sustained delivery of fundamental ecosystem services, such as the production of food, fuel, and shelter, the regulation of the quality and quantity of water supply, the control of natural hazards, etc. (see Figure 1, path 3).

Figure 1: Biodiversity Is Both a Response Variable Affected by Global Change Drivers and a Factor That Affects Human Well-Being. Links developed in this article are indicated. In the biodiversity box, the hierarchical components of biodiversity (genotypes, species, functional groups, and landscape units) each have the characteristics listed in the sub-box and explained in Figure 2 (number, relative abundance, composition, spatial distribution, and interactions involved in "vertical" diversity). Modified from [3,4].

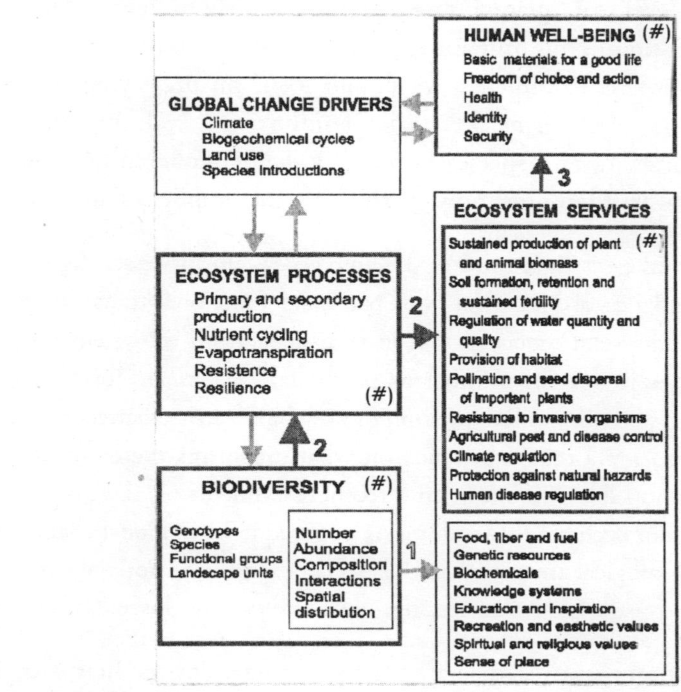

DOI: 10.1371/journal.pbio.0040277.g001

A few key messages can be drawn from existing theory and empirical studies. The first is that the number and strength of mechanistic connections between biodiversity and ecosystem processes and services clearly justify the protection of the biotic integrity of existing and restored ecosystems and its inclusion in the design of managed ecosystems.

All components of biodiversity, from genetic diversity to the spatial arrangement of landscape units, may play a role in the long-term provision of at least some ecosystem services.

However, some of these components are more important than others in influencing specific ecosystem services.

The evidence available indicates that it is functional composition—that is, the identity, abundance, and range of species traits—that appears to cause the effects of biodiversity on many ecosystem services. At least among species within the same trophic level (e.g., plants), rarer species are likely to have small effects at any given point in time. Thus, in natural systems, if we are to preserve the services that ecosystems provide to humans, we should focus on preserving or restoring their biotic integrity of terms of species composition, relative abundance, functional organization, and species numbers (whether inherently species-poor or species-rich), rather than on simply maximizing the number of species present.

Another key message is that, precisely because ecosystem processes depend on the presence and abundance of organisms with particular functional traits, there is wide variation in how ecosystem services—that in turn depend on ecosystem processes—respond to changes in species number as particular species are lost from or get established in the system. So, to the question of how biodiversity matters to ecosystem services, we have to reply that it depends on what organisms there are. Daunting? Certainly, but not hopeless. We know from recent assessments (1, 2, 7, 8) that global biodiversity loss is not occurring at random. As a consequence of global change drivers, such as climate, biological invasions, and especially land use, not only is the total number of species on the planet decreasing, but there are also losers and winners.

On average, the organisms that are losing out have longer lifespans, bigger bodies, poorer dispersal capacities, more specialized resource use, lower reproductive rates, and other traits that make them more susceptible to human activities such as nutrient loading, harvesting, and biomass removal by burning, livestock grazing, ploughing, clear-felling, etc. A small number of species with the opposite characteristics are becoming increasingly dominant around the world (Figure 3). Because there are well-established links between functional traits of locally abundant organisms and ecosystem processes, especially for plants (9-12), it may become possible to identify changes in ecosystem processes and in ecosystem services that depend on them under different biodiversity scenarios.

What We Do Not Know: Cascades, Surprises, and Megadiversity Hot-Spots

Some ecosystem services show a saturating relationship to species number—that is, the ecosystem-service response to additional species is large at low number of species and becomes asymptotic beyond a certain number of species. We seldom know what this threshold number is, but we suspect it differs among ecosystems, trophic levels, and services. The experimental evidence indicates that, in the case of primary production (e.g., for plantbased agricultural products), nutrient retention (which can reduce nutrient pollution and sustain production in the long-term), and resistance to invasions (which incur damage and control costs in agricultural and other settings) by temperate, herbaceous communities, responses often do not show further significant increases beyond about ten plant species per square meter (3, 13). But in order to achieve this number in a single square meter, a much higher number of species is needed at the landscape level (14). What about slow-growing natural communities, or communities that consist of plant species with more contrasting biology? What about communities that typically include many more species—for example, the megadiverse forest hotspots of the Amazon and Borneo, where species number can exceed 100 tree species per hectare (15)? To what extent are all those species essential for the maintenance of different ecosystem processes and services? Ecological theory (16) and traditional knowledge (17, 18) suggest that a large number of resident species per functional group, including those species that are rare, may act as 'insurance" that buffers ecosystem processes and their derived services in the face of changes in the physical and biological environment (e.g., precipitation, temperature, pathogens), but these ideas have yet to be tested experimentally, and no manipulative experiment has been performed in any megadiversity hotspot.

Most of the links between biodiversity and ecosystem services summarized in Table 1 emerged from theory and manipulative experiments, involved biodiversity within a single trophic level (usually plants), and operated mostly at the level of local communities. However, the most dramatic examples of effects of small changes in biodiversity on ecosystem services have occurred at the landscape level and have involved alterations of food-web diversity through indirect interactions and trophic cascades. Most of these have been "natural experiments," that is, the unintended consequence of intentional or accidental removal or addition of certain predator, pathogen, herbivore, or plant species to ecosystems. These "ecological surprises" usually involve disproportionately large, unexpected, irreversible, and

Figure 2: The Different Components of Biodiversity

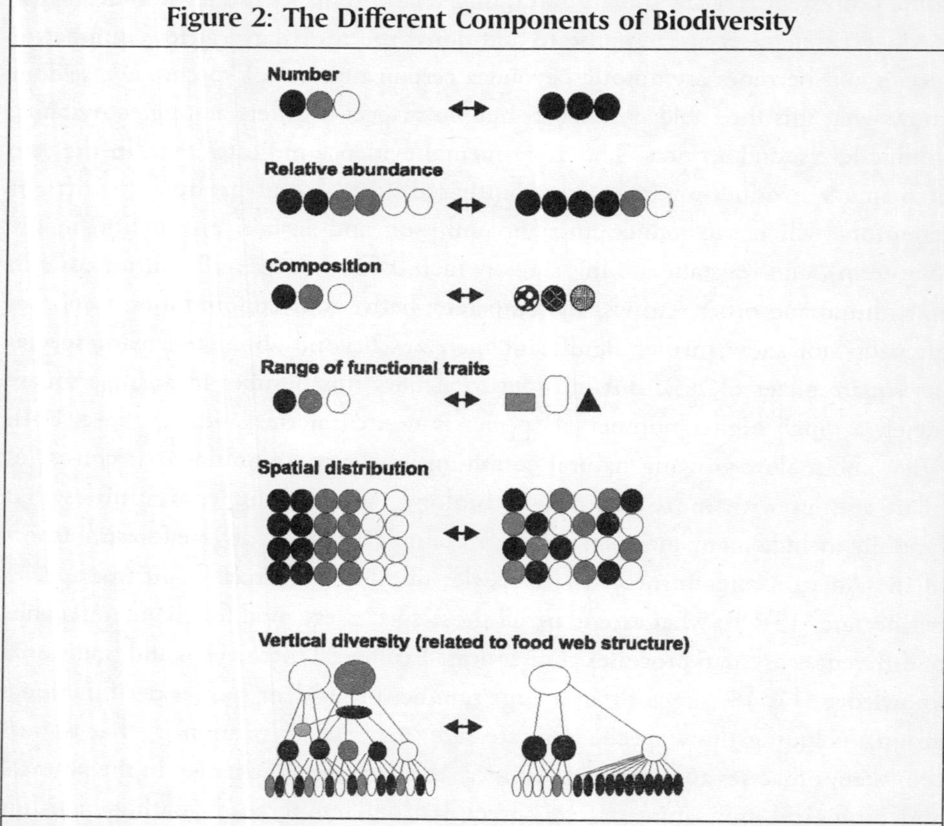

All of these components can be affected by human intervention (arrows), and in turn have repercussions for ecosystem properties and services. Symbols represent individuals or biomass units. Symbols of different shades represent different genotypes, phenotypes, or species.

negative alterations of ecosystem processes, often with repercussions at the level of ecosystem services, with large environmental, economic, and cultural losses. Examples include the cascading effects of decreases in sea otter population that led to coastal erosion in the North Pacific (19), and a marked decrease in grassland productivity and nutritional quality in the Aleutian islands as a consequence of decreased nutrientfiux from the sea by the introduction of Arctic foxes (20) (see (3) for a comprehensive list of examples). The vast literature on biological invasions and their ecological and socioeconomic impacts (21) further illustrates this point. Ecological surprises are difficult to predict, since they usually involve novel interactions among species. They most often result from introductions of predators, herbivores, pathogens and diseases, although cases involving introduced plants are also known. They do not depend linearly on species number or on well-established links between the functional traits of the species in question and putative ecosystem processes or services (3, 22).

Uneven Impacts: Biodiversity and Vulnerable Peoples

People who rely most directly on ecosystem services, such as subsistence farmers, the rural poor, and traditional societies, face the most serious and immediate risks from biodiversity loss. First, they are the ones who rely the most on the "safety net" provided by the biodiversity of natural ecosystems in terms of food security and sustained access to medicinal products, fuel, construction materials, and protection from natural hazards such as storms and floods (4). In many cases the provision of services to the most privileged sectors of society is subsidized but leaves the most vulnerable to pay most of the cost of biodiversity losses. These include, for example, subsistence farmers in the face of industrial agriculture (23) and subsistence. Shermen in the face of intensive commercial (Shing and aquaculture (24)). Second, because of their low economic and political power, the less privileged sectors cannot substitute purchased goods and services for the lost ecosystem benefis and they typically have little influence on national policy. When the quality of water deteriorates as a result of fertilizer and pesticide loading by industrial agriculture, the poor are unable to purchase safe water. When protein and vitamins from local sources, such as hunting and fruit, decrease as a result of habitat loss, the rich can still purchase them, whereas the poor cannot. When the capacity of natural ecosystems to buffer the effects of storms and floods is lost

Table 1: Biodiversity Components Affect Ecosystem Services in Multiple and Complex Ways

Ecosystem Services	Main Components of Diversity Involved and Mechanisms that Produce the Effect
Amount of biomass produced by plants considered important by humans	*** Functional composition—Faster-growing, bigger, more locally adapted plants produce more biomass, irrespective of the number of species present; in species-poor systems, coexisting plants with different resource use strategies or that facilitate each other's performance. ** Number of species—Within a constant resource and disturbance regime, a large species pool is more likely to contain groups of complementary or facilitating species and highly productive species, both of which could lead to higher productivity of the community.
Stability of biomass production by plants considered important by humans	*** Genetic diversity—Large genetic variability within a crop species buffers production against losses due to diseases and environmental change. *** Number of species—Cultivation of more than one species in the same plot or landscape maintains production over a broader range of conditions. *** Functional composition—Life history characteristics and resource use strategy of dominant plants determine the capacity of ecosystem processes to remain unchanged or return to their initial state in the face of perturbations.
Preservation of the fertility of soils that sustain the production of plants and animals considered important by humans	*** Functional composition—Fast-growing, nutrient-rich plants enhance soil fertility; dense root systems prevent soil erosion.
Regulation of quantity and quality of water available to humans, domestic animals, and crops	*** Arrangement and size of landscape units—Intact riparian corridors and extensive areas with dense vegetation cover reduce erosion and improve water quality. *** Functional composition—Vegetation dominated by large, fast-growing, big-leafed, deep-rooted plants has high transpiration rate, reducing stream flow.
Pollination essential for the immediate production of fruits by, and the perpetuation of, important plant species	*** Functional composition of pollinator assemblage—Loss of specialized pollinators leads to genetic impoverishment and lower number and quality of fruits.

Contd...

Contd...

	** Number of species of pollinator assemblage—Lower number of pollinator species leads to genetic impoverishment of plant species. ** Arrangement and size of landscape units—Large and/or well-connected landscape units allow movement of pollinators among plants of the same species, thus maintaining plant genetic pool.
Resistance to invasive organisms that have negative ecological, economic, and/or cultural impacts	*** Functional composition—Some key native species are very competitive or can act as biological controls to the spread of aliens. *** Arrangement of landscape units—Landscape corridors (e.g., roads, rivers, and extensive crops) can facilitate the spread of aliens; size and nature of suitable corridors are likely to be different for different organisms. ** Number of species—All else being equal, species-rich communities are more likely to contain highly competitive species and to contain less unused resources, and therefore be more resistant to invasions.
Pest and disease control in agricultural systems	*** Genetic diversity of crops—High intraspecific genetic diversity reduces density of hosts for specialist pests and, thus, their ability to spread. ** Number of crop, weed, and invertebrate species—High number of species acts similarly as genetic diversity and also increases habitat for natural enemies of pest species. ** Spatial distribution of landscape units—Natural vegetation patches intermingled with crops provide habitat for natural enemies of insect pests.
Regulation through biophysical feedbacks of climatic conditions suitable to humans and the animals and plants they consider important	*** Arrangement and size oflandscape units—Size and spatial arrangement of landscape units over large areas influence local-to-regional climate by lateral movement of air masses of different temperature and moisture; the threshold for effect is patch size of about 10 km diameter, depending on wind speed and topography. ** Functional composition—Height, structural diversity, architecture, and phenology modify albedo, heat absorption, and mechanical turbulence, thus changing local air temperature and circulation patterns.

Contd...

Contd...

Regulation through carbon sequestration in the biosphere of climatic conditions suitable to humans and the animals and plants they consider important	*** Arrangement and size of landscape units—Carbon loss is higher at forest edges, therefore as forest fragments decline in size or area/perimeter ratio, a larger proportion of the total landscape is losing carbon. ** Functional composition—Small, fast-growing, fast-decomposing, short-lived plants retain less carbon in their biomass than large, slow-growing, slow-decomposing, long-lived plants. * Number of species—High number of species can slow down the spread of pests and pathogens, which are important agents of carbon loss from ecosystems.
Protection against natural hazards (storms, floods, hurricanes, fires) that cause damage to humans and the animal production systems that they depend on	*** Arrangement and size of landscape units—Large patches of structurally complex vegetation or small, close-by patches are likely to offer more shelter to nearby ecosystems, and buffer them against flooding, sea intrusion, and wind. *** Functional composition—Deep-rooted plants are less susceptible to uprooting by hurricanes; extensive, mat-forming, superficial root systems protect soil against erosion by floods and storms; deciduous canopy types decrease flammability.

Asterisks indicate importance and/or degree or certainty (*** > ** > *) of the link between the ecosystem service in question and different components of biodiversity. Biodiversity components refer to plant assemblages unless otherwise specified. The putative mechanisms have been empirically tested in some cases, but remain speculative in others (modified from [3]). The list of ecosystem services is illustrative, rather than exhaustive.

Source: Modified from DOI: 10.1371/journal.pbio.0040277.t001

Figure 3: Lost Ecosystem Services and Vanishing Ecological Roles

Forest ecosystems in the tropics and subtropics are being quickly replaced by industrial crops and plantations. This provides large amounts of goods for national and international markets, but results in the loss of crucial ecosystem services mediated by ecological processes. In Argentina and Bolivia, the Chaco thorn forest (A) is being felled at a rate considered among the highest in the world (B), to give way to soybean cultivation (C), In Borneo, the Dypterocarp forest, one of the species-richest in the world (D) and the giant anteater in the Chaco plains (E), and the orangutan (F), is being replaced by oil palm plantations (G), These changes are irreversible for all practical purposes (H), Many animal and plant populations have been dramatically reduced by changing land use patterns, to the point that they could be considered functionally extinct, such as the maned wolf (I), and several species of pitcher plants (J), in the Bornean rainforest. Photos by Sandra Díaz, except (A and C), courtesy by Marcelo R. Zak.

because of coastal development (25), it is usually the people who cannot flee—for example, subsistence fishermen—who suffer the most. In summary, the loss of biodiversity-dependent ecosystem services is likely to accentuate inequality and marginalization of the most vulnerable sectors of society, by reducing their freedom of choice and action. Economic development that does not consider effects on these ecosystem services may decrease the quality of life of these vulnerable populations, even if other segments of society benefit. Biodiversity change is therefore inextricably linked to poverty, the largest threat to the future of humanity identified by the United Nations. This is a sobering conclusion for those who argue that biodiversity is simply an intellectual preoccupation of those whose basic needs and aspirations are fulfilled.

Future Directions

Most of the concrete actions to slow down biodiversity loss fall under the domain of policy making by governments and the civil society. However, the scientific community still needs to fill crucial knowledge gaps. First, we need to know more about the links between biodiversity and ecosystem services in species-rich ecosystems dominated by long-lived plants. Second, if we are to anticipate and avoid undesirable ecological surprises, better models and more empirical evidence are needed on the links between ecosystem services and interactions among different trophic levels. Third, we need to reinforce the systematic screening for functional traits of organisms likely to have ecosystem-level consequences. In this sense, our knowledge of how the presence and local abundance of organisms (especially plants) bearing certain attributes affect ecosystem processes has made considerable progress in the past few years. However, we know much less of how the range of responses to environmental change among species affecting the same ecosystem function contributes to the preservation of ecosystem processes and services in the face of environmental change and uncertainty (16, 26). This is directly relevant to risk assessment of the sustained provision of ecosystem services. Fourth, experimental designs for studying links between biodiversity and ecosystem processes and services need to not only meet statistical criteria but also mimic biotic configurations that appear in real ecosystems as a result of common land-use practices (e.g., primary forest versus monospecific plantations versus enrichment planting, or grazing-timber agroforestry systems versus a diverse grazing megafauna versus a single grazer such as cattle). In pursuing this, traditional knowledge systems and common management practices provide a valuable source of inspiration to develop new designs and testable hypotheses (27, 28). Finally, in order to assist policy decisions and negotiation among different local, national, and international stakeholders, considerable advance is needed in the evaluation and accounting of ecosystem services (29, 30). The challenge here is to find ways to identify and monitor services that are as concrete as possible, but at the same time not alienate the view of less powerful social actors or bias the analysis against services that are difficult to quantify or grasp.

The Bottom Line

By affecting the magnitude, pace, and temporal continuity by which energy and materials are circulated through ecosystems, biodiversity in the broad sense influences the provision of ecosystem services. The most dramatic changes in ecosystem services are likely to come from altered functional compositions of communities and from the loss, within the same trophic level, of locally abundant species rather than from the loss of already rare species. Based on the available evidence, we cannot define a level of biodiversity loss that is safe, and we still do not have satisfactory models to account for ecological surprises. Direct effects of drivers of biodiversity loss (eutrophication, burning, soil erosion and flooding, etc.) on ecosystem processes and services are often more dramatic than those mediated by biodiversity change. Nevertheless, there is compelling evidence that the tapestry of life, rather than responding passively to global environmental change, actively mediates changes in the Earth's life-support systems. Its degradation is threatening the fulfilment of basic needs and aspiration of humanity as a whole, but especially, and most immediately, those of the most disadvantaged segments of society.

Acknowledgements

We are grateful to W Reid, H A Mooney, G Orians, and S Lavorel for encouragement, inspiration, and critical comments during the process that led to this article, and to the leading authors of Millennium Ecosystem Assessment's Current State and Trends, chapter 11.

(Sandra Díaz is principal researcher and associate professor of ecology and biogeography at Instituto Multidisciplinario de Biología Vegetal (CONICET-UNC)

and FCEFyN, Universidad Nacional de Argentina. Joseph Fargione is research assistant faculty at the Department of Biology, University of New Mexico, Albuquerque, New Mexico, United States of America. F Stuart Chapin III is professor of ecology at the Institute of Arctic Biology, University of Alaska at Fairbanks, Fairbanks, Alaska, United States of America. David Tilman is the McKnight Presidential Chair in Ecology at the Department of Ecology, Evolution and Behavior, University of Minnesota, St. Paul, Minnesota, United States of America. To whom correspondence should be addressed. The autours can be reached at sdiaz@com.uncor.edu).

References

1. Baillie JEM, Hilton-Taylor C, Stuart SN (2004), IUCN Red List of Threatened Species: A Global Species Assessment. Gland (Switzerland): IUCN.
2. Mace G, Masundire H, Baillie J, Ricketts T, Brooks T, *et al.* (2005), Biodiversity. In: Hassan R, Scholes R, Ash N, editors. Ecosystems and human well-being: Current state and trends: Findings of the Condition and Trends Working Group. Washington (D. C.): Island Press. pp. 77–122.
3. Díaz S, Tilman D, Fargione J, Chapin FI, Dirzo R, *et al.* (2005), Biodiversity regulation of ecosystem services. In: Hassan R, Scholes R, Ash N, editors. Ecosystems and human wellbeing: Current state and trends: Findings of the Condition and Trends Working Group. Washington (D. C.): Island Press. pp. 297–329.
4. Millennium Ecosystem Assessment (2005) Ecosystems and human well-being: Biodiversity synthesis. Washington (D. C.): World Resources Institute. 86 p.
5. Stokstad E (2005), Ecology: Taking the pulse of earth's life-support systems. Science 308: 41–43.
6. Kremen C (2005), Managing ecosystem services: What do we need to know about their ecology? Ecol Lett 8: 468–479.
7. Kotiaho JS, Kaitala V, Komonen A, Paivinen J (2005), Predicting the risk of extinction from shared ecological characteristics. Proc Natl Acad Sci USA 102: 1963–1967.
8. McKinney M, Lockwood J (1999), Biotic homogenization: A few winners replacing many losers in the next mass extinction. Trends Ecol Evol 14: 450–453.
9. Grime JP (2001), Plant strategies, vegetation processes, and ecosystem properties. Chichester (United Kingdom); New York: John Wiley & Sons. 417 p.
10. Eviner VT, Chapin FS (2003), Functional matrix: A conceptual framework for predicting multiple plant effects on ecosystem processes. In: Futuyma DJ, editor. Annual review of ecology evolution and systematics, volume 34. Palo Alto (California): Annual Reviews. pp. 455–485.

11. Díaz S, Hodgson JG, Thompson K, Cabido M, Cornelissen JHC, *et al.* (2004), The plant traits that drive ecosystems: Evidence from three continents. J Veg Sci 15: 295–304.

12. Garnier E, Cortez J, Billès G, Navas ML, Roumet C, *et al.* (2004), Plant functional markers capture ecosystem properties during secondary succession. Ecology 85: 2630–2637.

13. Hooper DU, Chapin FS, Ewel JJ, Hector A, Inchausti P, *et al.* (2005), Effects of biodiversity on ecosystem functioning: A consensus of current knowledge. Ecol Monogr 75: 3–35.

14. Tilman D (1999), Diversity and production in European grasslands. Science 286: 1099–1100.

15. Phillips OL, Hall P, Gentry AH, Sawyer SA, Vasquez R (1994), Dynamics and species richness of tropical rain-forests. Proc Natl Acad Sci USA 91: 2805–2809.

16. Elmqvist T, Folke C, Nystrom M, Peterson G, Bengtsson J, *et al.* (2003), Response diversity, ecosystem change, and resilience. Frontiers in Ecology and the Environment 1: 488–494.

17. Trenbath B (1999), Multispecies cropping systems in India: Predictions of their productivity, stability, resilience and ecological sustainability. Agroforestry Systems 45: 81– 107.

18. Altieri M (2004), Linking ecologists and traditional farmers in the search for sustainable agriculture. Frontiers in Ecology and the Environment 2: 35–42.

19. Estes JA, Tinker MT, Williams TM, Doak DF (1998), Killer whale predation on sea otters linking oceanic and nearshore ecosystems. Science 282: 473–476.

20. Maron JL, Estes JA, Croll DA, Danner EM, Elmendorf SC, *et al.* (2006) An introduced predator alters Aleutian Island plant communities by thwarting nutrient subsidies. Ecol Monogr 76: 3–24.

21. Mooney HA, Mack RN, McNeely J, Neville LE, Schei PJ, *et al.* (2005), Invasive alien species: A new synthesis. Washington (D. C.): Island Press. 368 p.

22. Walker B, Meyers JA (2004), Thresholds in ecological and social-ecological systems: A developing database. Ecology and Society 9. Available: *http://www.ecologyandsociety.org/vol9/iss2/art3*. Accessed 23 June 2006.

23. Lambin EF, Geist HJ, Lepers E (2003), Dynamics of land-use and land-cover change in tropical regions. Annual Review of Environment and Resources 28: 205–241.

24. Naylor RL, Goldburg RJ, Primavera JH, Kautsky N, Beveridge MCM, *et al.* (2000), Effect of aquaculture on worldfish supplies. Nature 405: 1017–1024.

25. Danielsen F, Sorensen MK, Olwig MF, Selvam V, Parish F, *et al.* (2005), The Asian tsunami: A protective role for coastal vegetation. Science 310: 643.

26. Lavorel S, Garnier E (2002), Predicting changes in community composition and ecosystem functioning from plant traits: Revisiting the Holy Grail. Funct Ecol 16: 545–556.

27. Díaz S, Symstad AJ, Chapin FS, Wardle DA, Huenneke LF (2003), Functional diversity revealed by removal experiments. Trends Ecol Evol 18: 140–146.

28. Scherer-Lorenzen M, Potvin C, Koricheva J, Bornik Z, Hector A, *et al.* (2005), The design of experimental tree plantations for functional biodiversity research. In: Scherer-Lorenzen M, Körner C, Schulze ED, editors. The functional significance of forest diversity. Berlin: Springer-Verlag. pp. 377–389.

29. DeFries R, Pagiola S, Adamowicz W, Resit Akçakaya H, Arcenas A, *et al.* (2005), Analytical approaches for assessing ecosystem conditions and human well-being. In: Hassan R, Scholes R, Ash N, editors. Ecosystems and human well-being Current state and trends: Findings of the Condition and Trends Working Group. Washington (D C.): Island Press. pp. 37–71.

30. Boyd J, Banzhaf S (2006), What are ecosystem services? The need for standardized environmental accounting units. Washington (D C): Resources for the Future.

4

EPOC High-Level Special Session on the Costs of Inaction
The Costs of Inaction with Respect to Biodiversity Loss
Background Paper

Geoffrey Heal

Biodiversity is required for the functioning of many ecosystems and biogeochemical cycles that are essential to human well-being. Loss of biodiversity implies the loss of the services of these systems to the detriment of human welfare. Policy inaction increases the total loss of biodiversity and the services that are dependent on it. Although we know that biodiversity is important, we do not know exactly how important, nor the precise extent of the welfare costs imposed by its loss. This uncertainty, coupled with the irreversibility of biodiversity loss and the possibility of learning more about its value, implies a real option value to the conservation of biodiversity and a rationale for a precautionary approach. While the overall cost of biodiversity loss is unknown, some parts of this cost can be estimated, including the costs of lost bioprospecting leads, the

Source: Geoffrey Heal (Consultant) Epoc High-Level Special Session on the Costs of Inaction The Costs of Inaction with Respect to Biodiversity Loos Background Paper, www.oecd.org/dataoecd/37/3/34738405.pdf

costs of lost carbon storage, the costs of lost tourism business and the costs of diminished watershed protection. These very partial costs amount to many tens of billions of dollars.

1. Why Biodiversity Loss Matters

The welfare of human societies depend on the functioning of a range of diverse natural ecosystems, which provide critical and often irreplaceable ecosystem services. Many authors have discussed and documented this dependence (Daily 1997, Ecological Society of America 1997, Millennium Ecosystem Assessment.) Natural ecosystems are responsible for services as diverse as climate stabilization, crop pollination, soil fertility, waste disposal, water purification, flood control, pest control, recreational services, and many others.

Biodiversity is a critical ingredient of most natural ecosystems: while the exact nature of the relationship between biodiversity and ecosystem function is still a research topic (Grime 1997, Hooper and Vitousek 1997, McGrady-Steed 1997, Naeem and Li 1997), there is general agreement that loss of biodiversity weakens ecosystems, making them less resilient and less productive, and reduces their capacity to provide the services needed by humans. Tilman's (1997) study showed that the average amount of biomass grown per year on a plot of a given size increased with the diversity of functional groups represented. The increase leveled off after a certain point, above which more diversity adds little to the community's performance. Tilman and co-workers also found more nutrient uptake and better soil quality on plots with a more diverse collection of plant species. Furthermore, the plots that were more diverse in this sense were also more robust in the face of weather fluctuations. It appears from this work that both functional diversity and species diversity are important in maintaining productivity and resilience.

There are several different ways of measuring biodiversity (see Armsworth *et al.*, 2004 and Nunes *et al.*, 2001 for details and for comments on how the measurement of biodiversity relates to economic valuation) but the differences are not significant for the discussions in this paper. We can take biodiversity to be of some measure of the total number of species, which may be a global measure or a local one. Locally the loss of biodiversity in a specific ecosystem may weaken that system and make it less productive, so that localized losses of biodiversity may lead to a loss of human welfare even if not a part of a global biodiversity loss. A global loss will certainly mean that some ecosystems are weakened and their services reduced. There is also general agreement that the current rate of loss of biodiversity is unprecedentedly high, both for populations (i.e., locally) and for species (i.e., globally) (Hughes *et al.*, 1997). An implication of this is that biodiversity loss threatens human welfare through the reduction of the ecosystem services upon which we depend.

There are other mechanisms through which biodiversity loss threatens human welfare. Biodiversity has made, and indeed continues to make, a vast contribution to agricultural productivity. It is no exaggeration to say that the green revolution of the second half of the twentieth century allowed us to feed a doubled human population because of clever use of biodiversity, and with the prospect of at least four billion more sharing the earth's bounty within decades there is clearly a need for continued use of biodiversity in this way. Plant breeders find use for genetic diversity within a species by developing varieties resistant to disease, drawing on the fact that different varieties of the same species have different degrees of susceptibility to any particular disease. If one were to plant a single variety and a disease to which it is susceptible were to strike, the entire planting would be destroyed. If instead one plants different varieties, which typically differ in their susceptibility to a disease, then there is some insurance against complete crop loss (for a discussion of whether agricultural biodiversity is adequately used for this purpose see Heal *et al.*, 2004). The Irish potato famine of the nineteenth century is an example of the extreme hardship caused by growing a single variety of potato, Solanum tuberosum. So genetic variation at the species level is economically valuable and has historically been the source of almost all agricultural progress.

A dramatic example of the direct importance of diversity to humans comes from the recent history of rice production. The prosperity and comfort of literally billions of people depends on rice harvests. In the 1970s, the grassy stunt virus, a new virus carried by the brown plant hopper, threatened the Asian rice crop. The virus appeared capable of destroying a large fraction of the crop: in some years it destroyed as much as one-quarter. Developing a form of rice resistant to this virus became of critical importance. Rice breeders succeeded in this task with the help of the International Rice Research Institute (IRRI) in the Philippines. The IRRI conducts research on rice production and holds a large seed bank of about 80,000 different varieties of rice and the near-relatives of rice. In this case the IRRI located a single strain of wild rice that was not used commercially and was resistant to the grassy stunt virus. The gene conveying resistance was transferred to commercial rice varieties, yielding commercial rice resistant to the threatening virus. This would not have been possible without the genes from a strain of rice that was apparently of no commercial value: without this variety, the world's rice crop, one of its most important food crops, would have been seriously damaged by the new virus. The strain of wild rice that was resistant to the virus was found in only one location; a valley that was flooded by a hydroelectric dam shortly after the IRRI found and took into its collection the critical rice variety. The same situation occurred later in the 1970s, and similar stories have occurred with other food crops, in particular corn in the United States (Myers 1997).

Biodiversity also contributes to the development of human knowledge and understanding, especially in the medical arena. Biodiversity's store of genetic material is a priceless source of knowledge. A good example is provided by the Polymerase Chain Reaction (PCR), a reaction central to the amplification of DNA specimens for analysis, as in forensic tests used in criminal investigations and many processes central to the biotechnology industry. Culturing – the process of taking a minute sample of DNA and multiplying it manifold – requires an enzyme that is resistant to high temperatures. Enzymes with the right degree of temperature resistance were found in organisms in the hot springs in Yellowstone National Park and used to develop an enzyme for culturing DNA specimens. This enzyme is now central to the rapidly growing biotechnology industry and the pharmaceuticals that it is producing. Other widely-cited illustrations of the use of biodiversity in producing pharmaceuticals are the extraction of taxol, a treatment for breast cancer, from willow bark, and the extraction of vincristine and vinbalstine, treatments for childhood leukemia, from the rosy periwinkle.

These examples clearly show that human welfare is threatened by the loss of biodiversity. As biodiversity loss is a continuing process, a part of the cost of policy inaction is that the longer we wait to stem this loss, the greater the final total cost to human societies. And it is important to note that biodiversity loss is in general irreversible – extinction of a species is a collective death of that species and a collective transit to "that undiscovered bourne from which no traveler has yet returned." So the longer we wait to act, the greater the costs we will ultimately pay.

2. Uncertainties Concerning Biodiversity Loss

However, the problem is in fact far more complex than this analysis suggests. Biodiversity loss (conceived for the moment as extinction of species and extinction of populations) is not always easily measured, nor is it easy to say how human actions – which are the predominant cause of biodiversity loss – enter into this process. Biodiversity loss usually follows human actions with a long and unpredictable lag. And of course some elements of biodiversity matter more to us than others. All of these issues make assessment of the costs of policy inaction considerably more complex than they might appear at first sight.

There is much uncertainty about the rate of biodiversity loss, even though all parties to the debate agree that the rate of loss is high, higher than at any time in the last several thousand years. Uncertainty comes in part from the fact that we do not really know how many species there are, which introduces the possibility that we may be losing species of whose existence we are unaware. It also comes from the fact that we have only limited information about the populations of many species, and so are not well-placed to estimate the threats to their continued existence.

But perhaps more important than this statistical uncertainty about populations and their rates of change is a lack of knowledge of how human activities impact a species' ability to continue. The main causes of species loss are deforestation and the clearing of land for agriculture – some of which is included in deforestation. Climate change will probably join them within the next decades. Many ecosystems respond in a very nonlinear fashion to stress: they can cope with outside stress with little or no damage up to a point, but are damaged substantially if the stress passes a critical level. A possible illustration of this is the species-area relationship, which finds that the number of species S on an area of land A increases with a fractional power of the area: $S = kAc$ where k and c are constants, $k > 0$ and $0 < c < 1$ (Ehrlich and Roughgarden 1987). Plotting the number of species vertically against area horizontally therefore gives a curve that is steep near the origin and flatter further out. So cutting back the area available for species from a large level, we see that initially a given reduction in area leads to only a small loss in the number of species, whereas once the area has been reduced substantially the same reduction in area leads to a far larger drop in species and in biodiversity. There are other more dramatic examples of ecosystems responding nonlinearly to human stresses, perhaps the best of which are the classic studies of eutrophication of freshwater lakes by Carpenter *et al.* (1999). Here a very small change in the phosphorus loading of a lake can cause it to flip from a pure and productive to a eutrophic state, with little or no warning and serious economic consequences to lake users.

Time lags in responses add a further element of uncertainty here. Biologists have a concept of the minimum viable population, the smallest population of a species that has a high probability of continuing in existence (Ehrlich and Roughgarden 1987). Once the population is below this level the species is effectively extinct, in the sense that it will eventually die out. But if the life of members of the species is long, it might be decades before actual extinction occurs and there is agreement that this has happened. For example, macaws live for sixty years or more: the populations of several species are thought to be below the minimum viable level today (Beissinger and Snyder, 1991), but as there are breeding young adults alive, these species will still be visible, albeit on a reduced scale, for perhaps as long as the next century. Biologists refer to such species rather graphically as the "walking dead."

Figure 1: Relationship between Species Habitat Area and Number of Species

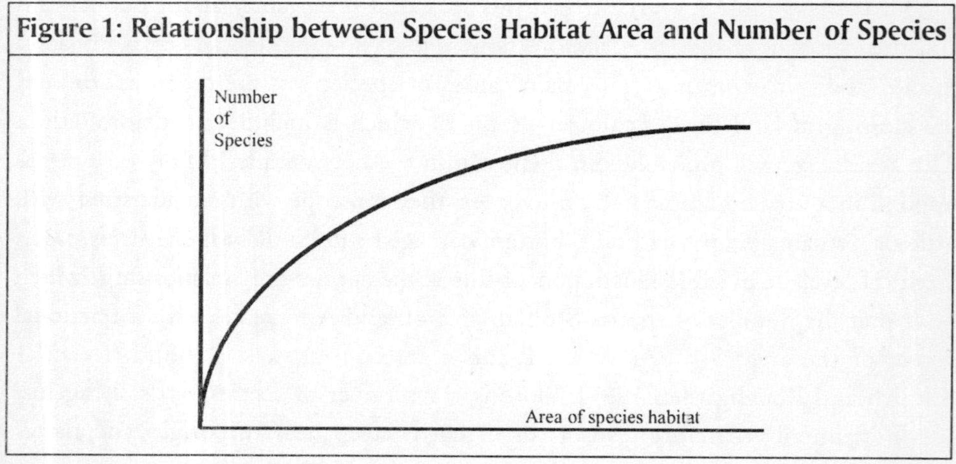

These uncertainties and nonlinearities complicate greatly the assessment of the costs of policy inaction. In the simplest of all worlds, where everything is known, calculating the cost of delaying intervention in the process of biodiversity loss would also be simple. We would know that a specific measure, such as incentives for the conservation of tropical forests, would cut the rate of species loss from X per year to Y < X per year. Delaying introducing the measure for T years would therefore cause the loss of T(X-Y) extra species. To go further we would have to know the welfare losses associated with species loss: suppose for the moment that the loss of a species imposes a welfare loss whose present value is $L. Then the cost of T years of policy inaction is $LT(X-Y).

In practice the only number we know in this expression is T, the time for which we are delaying the policy. We do not know the current rate of biodiversity loss X, nor the rate to which this could be reduced by an intervention such as cutting deforestation, Y. And we do not know the value of a species either, although I will review below some estimates.

3. The Timing of Interventions

A common response to such extensive uncertainty is to postpone a decision, in the hope that we will learn more about some of the currently-unknown parameters – the Xs, Ys and Ls above. Then on the basis of better scientific evidence we will be able to proceed with a more rational and defensible choice, always attractive to decision-makers. But we have to recognize that there is a cost to waiting for more information, the cost of the biodiversity that is lost in the meantime, a cost that is unknown. In fact there are some important and far-from-obvious analytical issues raised by postponing decisions in order to wait for better information, which are captured under the title of "real option values." (For a detailed discussion, see Dasgupta and Heal, 1979.)

The point associated with option values is as follows. If you have the option of destroying something irreplaceable now, or postponing the destruction in the hope of learning more about how important it is and what the consequences of loss would be, then in comparing the two options the choice to postpone destruction and conserve has to be credited with an "option value," a value which arises because if you conserve the item then the decision about whether to destroy can be revisited again later in the light of additional and better information. So by postponing destruction you may be able to make better decisions later on, and the expected gain from this constitutes the option value. It is the value of retaining an option for the choice to go either way, an option that is foreclosed if an irreplaceable object is destroyed. Roughly it is the value of taking a flexible position that keeps options open. From this it follows that conserving biodiversity has to be credited with an option value, in addition to any direct benefits expected to be associated with the conservation. Very few cost-benefit analyses of biodiversity conservation have taken this into consideration. (There are estimates of option values associated with conservation of forests in North America, where the values of the forest are largely a function of its timber and recreational values, and there is the possibility of developing the land-see Conrad 2000 and Buttle and Rondeau 2004). If policy intervention is postponed and more biodiversity is lost, then this option value is also lost.

. A natural extension of the observation that better decisions can be made if one waits for additional information is the use of adaptive management, which is a relatively new paradigm for confronting the uncertainty amongst management policy alternatives for large complex ecosystems or ecosystems where functional relationships are poorly known (Walters, 1986; Holling, 1973). A key component of adaptive management is active learning by introducing new management policies to learn more about the system's behavior and so reduce uncertainty. Typically, there may be an effort to implement environmental management actions as "experiments" in order to "learn by doing," with the experiments designed to reduce the critical uncertainties about the ecosystem's behavior. Adaptive management thus provides for a mechanism for learning systematically about the links between human societies and ecosystems. In contrast, the learning that occurs in economic models with option values is purely passive—information about the value of an environmental system is acquired with the passage of time. Adaptive management is a natural step from the passive concept of an option value associated with gaining information to the concept of managing the ecosystem to learn and so reduce uncertainty. When an adaptive management approach is possible, the option value associated with conservation is likely to be increased because of the enhanced rate of information acquisition.

4. Risk Aversion

Given all the uncertainties associated with the value of biodiversity conservation, one may be tempted to invoke the precautionary principle. Notably, the 1992 Rio Declaration (Article 15) (see Gollier *et al.*, 2000) stated: "where there are threats of serious and irreversible damage, lack of full scientific certainty shall not be used as a reason for postponing cost-effective measures to prevent environmental degradation." Though the precautionary principle has been attacked as being a vague concept lacking a precise definition, the essence of the precautionary principle is clear[1] and is that the burden of proof should be to demonstrate that changes do not cause irreversible environmental damage, rather than proving a change is dangerous. Biodiversity loss is clearly a serious form of irreversible damage.

Most economists, if asked to think of a justification for the precautionary principle in decision-making, would probably couch it in terms of learning, irreversibilities, and option values. The option value linked to conserving an ecosystem whose change is irreversible is in effect a reward for cautious behavior, although it certainly does not imply that conservation is always appropriate. Gollier *et al.* (2000) note that the precautionary principle can also be given a formal justification in environmental decision-making without invoking irreversibilities, just assuming that there is cumulative damage from human actions and possible learning over time about the consequences of the damage.

The unpredictability of the outcome of an environmental policy under uncertainty means that while the outcome could be excellent, it also has a chance of being poor. In general, faced with the choice between policies that generate the same expected value but with different ranges of outcomes, most people would choose the policy with the lowest variability. Risk aversion is a measure of what a person is willing to pay to avoid a risk and replace it by a certain outcome. If people are very risk-averse, then an environmental policy that delivers a modest outcome with some certainty might be preferred to one that may deliver a truly outstanding outcome but may also deliver a very poor result. In such situations, a policy maker has to decide whether to build some measure of risk aversion into the analysis, and if so how much. There are studies of the degree of risk-aversion displayed by individuals in financial markets (see Chetty 2003 and the references therein), but as risk aversion for a given person may vary with the magnitude of the risk and as it varies across people, these are not necessarily the appropriate values to use in environmental studies.

If society is extremely risk averse, then the objective of maximizing the expected value of society's welfare can be replaced by an objective known as "maximin." The intent in such cases is to focus on the worst possible outcome, the minimum, and then seek the policy intervention that maximises the welfare gains in the event that this outcome arises (hence the name: for a discussion see Arrow and Hurwicz 1972 and Maskin 1979). This policy will be different from that which would have been adopted on the basis of certainty equivalence. By way of illustration, consider an aquatic ecosystem that amongst others provides flood control services to a residential area. It is possible that decision-makers believe that the loss of human life through floods is the worst possible outcome and must be prevented at all costs. Such a belief would be appropriately represented by maximin preferences which would lead the analyst to select as best the project that minimizes the loss of life from flooding. Focusing exclusively on the worst possible outcome is only justified if there are good reasons to suppose that society is really risk averse and is willing to sacrifice considerable upside potential of a policy to avoid any chance of a bad outcome. Technically, the maximin objective can be seen as a limiting case of the expected welfare objective as the degree of risk aversion increases without limit. There are also arguments that suggest that the maximin may be an appropriate choice of objective in some cases of ambiguity, that is, cases in which there are no objective or subjective probabilities (Arrow and Hurwicz 1972 and Maskin 1979). Intuitively there is some connection between the precautionary principle and the maximin approach: both emphasize the worst possible outcome and seek to avoid it. Both would suggest placing great priority on stemming the loss of biodiversity because of the possibility of very costly losses that cannot currently be anticipated.

5. The Costs of Biodiversity Loss

We have seen that there are many complications in making rational decisions about the conservation of biodiversity. We do not know exactly how much we have nor how much we are losing. Nor do we know exactly how the rate of loss depends on our actions; although we do know the general qualitative properties of this relationship – we know which of our actions are driving biodiversity loss and how to reduce that loss. Are there any solid economic grounds on which we can base our analysis? At the root of our difficulties in valuing biodiversity is the fact that many of the services that it provides are public goods – for example, climate stabilization and genetic knowledge – so that in many cases there is no market for them and so no market price. When there is a market for biodiversity-based services, as with ecotourism based on charismatic megafauna (see section 5.4 below) then the market captures only a part of society's willingness to pay for the service (Heal 2003); so that the market undervalues biodiversity even when it does value it. Furthermore we can rarely value biodiversity directly: we value the services of the ecosystems of which biodiversity is an integral and essential component. In other words, biodiversity leads to services, which produce value: we work back to attribute some value to the biodiversity. Without the underlying biodiversity there would be no services and so no value. This does not necessarily imply that all the biodiversity in the underlying ecosystems is needed to produce the services, but it is the general opinion of systems ecologists that few if any components of an ecosystem are redundant (see Chapin *et al.* 2000). In this case it is reasonable to attribute the end value to the underlying biodiversity.

There are some estimates of the values of some of the ecosystem services provided by systems that depend on biodiversity and which are destroyed in the process of biodiversity loss. These are estimates of the values of biodiversity as a source of genetic information for use in pharmaceutical research, the value of forests in carbon sequestration, the value of biodiversity in watersheds as water purification systems and the value of charismatic fauna as tourist destinations.[2] All of these have been reasonably well-established. Unfortunately they cover only a small set of the values of services of biodiversity in total and so can only be taken as a lower bound on the overall value. Nevertheless they are high enough to give a sense of the economic importance of biodiversity loss. (For an earlier review of some of these values, and an excellent overview of the issues raised by biodiversity valuation, see Pearce 2001).

5.1 Biodiversity and Bioprospecting

Bioprospecting refers to the use of naturally-occurring substances as potential sources of pharmacologically active and valuable compounds, as what the pharmaceutical industry call "leads." Over one-third by value of prescription drugs originated as naturally-occurring compounds, so bioprospecting has been an important source of biochemical leads in drug development. Several drug companies have signed agreements with tropical countries under which they will systematically investigate the potential of compounds derived from plants or insects, and sometimes from vertebrates such as snakes, as pharmacological leads (Heal, 2000). Typically such agreements involve an up-front payment by the company and then royalties on the drugs developed. This has lead to a number of studies of the value that might be assigned to biodiversity in this process, the most recent of which is Rausser and Small (2000) (see also Simpson, Sedjo and Reid 1996, and for a review of the economic characteristics of biodiversity in the context of genetic resources see OECD 2001). Rausser and Small look at the value of biodiversity in known biodiversity "hot spots" from the perspective of drug development, and conclude that this could be worth as much as USD 9,000 per hectare. It bears emphasizing that this is an upper limit and applies only to the areas in which biodiversity is most concentrated. And if two different areas contain similar biodiversity, we cannot sum their values as this would involve some double-counting.

We can use this to put a dollar value on the loss of some of the tropical rainforests, losses that could be halted by policy interventions that are now available. Many millions of hectares of tropical forest are lost each year. Not all of this is in biodiversity "hot spots," however: a guess is that one million hectares of biodiversity-rich forests are lost each year. And not all of these contain unique biodiversity. Nevertheless, it is possible that the annual loss of biodiversity due to deforestation is costing humanity billions of dollars annually solely from the perspective of new drug development, that is, each year we are losing new drug leads that could be worth billions of dollars.

5.2 Biodiversity and Carbon Sequestration

Biodiversity is an integral part of ecosystems, often of complex ecosystems. Nowhere is this more true than with tropical forests, ecosystems whose components are a vast range of species that constitute a significant fraction of Earth's biodiversity. These systems draw carbon from the atmosphere and store it above and below ground. The Kyoto Protocol establishes emissions markets which put a price on carbon emissions and on carbon sequestration. Estimates of what this might be if the Kyoto Protocol were fully implemented run in the range of USD 20 to USD 50 per ton of CO2. More concretely, the European Union's current emissions trading scheme has that price at about EUR 10 per ton of CO2. Forests may sequester carbon: in fact the terrestrial biosphere is thought to sequester about 30% of the CO2 emitted by the combustion of fossil fuels. Forests sequester primarily when they are growing: growing moist tropical forests may sequester as much as 20 tons of CO2 per hectare per year, an ecosystem service worth EUR 200 per hectare per year at current European Union Emissions Trading Scheme (ETS) prices and considerably more at the prices associated with full implementation of the Kyoto Protocol. The capitalized value of such a flow of services is several thousand Euros per hectare, which is additional to any value that the forest may have for bioprospecting.

Climax forests, forests that are in an equilibrium age distribution of trees, sequester much less. These are the forests which are extensively logged and of which a significant percentage is cut each year. The cutting of these forests releases carbon dioxide back into the atmosphere – the last IPCC reports estimate that between 20% and 25% of all CO2 released into the atmosphere in the 1990s came from deforestation. This makes deforestation as important a source of heat-trapping gases as the use of fossil fuels in the US. Tropical forests cover about 2.7 billion hectares, and about 15% are cut down each year. That makes an annual loss of tropical forests of about 200 million hectares. If each hectare of deforestation releases Q tons of CO2 and we take the current EU ETS price for CO2, then the total value of this CO2 release from deforestation is 10x200,000,000xQ. As the number Q could be about 10, this is a value that is again in the tens of billions. So the conservation of even climax forests, which are not large carbon sinks but are large stocks of carbon, could be providing carbon storage services worth tens of billions. Climax tropical forests constitute much of the world's biodiversity. Again, this is additional to any bioprospecting value that this may have.[3]

5.3 Biodiversity and Watersheds

Watersheds are some of the most valuable of all ecosystems, as they fulfill two crucial functions – the stabilization of stream flows and the purification of the water that flows through (National Academy of Sciences, 2000). As water is a critical human need and an increasingly scarce commodity, these functions are of signal importance. Their economic significance is well illustrated by the choice that New York City made in 1997 to invest USD 1.5 billion in the restoration of its Catskill watershed, as a way of avoiding the need to introduce water filtration plants.

Watersheds are a specific type of ecosystem, and it is the biodiversity in them that allows them to purify water and control stream flow. Specifically, as rain falls into a watershed and percolates through the soil, microorganisms in the soil remove impurities from the water, and the slow flow through the soil facilitates sedimentation of additional impurities. In New York's case, development in the neighbourhood of the watershed had overloaded its capacity to process impurities, and consequently the City's response was to invest in upgrading sewage systems, buying undeveloped land to prevent further development, and pay farmers in the region to maintain riparian buffer zones and to adopt organic agriculture. (For details see National Academy of Sciences, 2000.) Similar measures have now been adopted in many other watersheds – for example, the owners of the Perrier brand pay local farmers to adopt organic farming methods to avoid compromising the watershed. So the biodiversity in the various watersheds around the world is worth many billions of dollars.

5.4 Biodiversity and Tourism

When the general public thinks of biodiversity, it generally thinks of what conservation biologists term charismatic megafauna, exemplified by lions and tigers and elephants and giraffes and the great herds of ungulates on the African plains. That these have economic value is shown by people's revealed willingness to pay substantial amounts to view them and spend time in their presence. For many African countries (including Botswana, Namibia, South Africa, Zambia and Kenya), ecotourism based on their endowments of charismatic megafauna is one of the major contributors to national income, and the support of this aspect of biodiversity often is the highest value use of land. In addition to providing income, ecotourism has the attractive feature that it generates income and employment in remote rural areas and so contributes to rural development and the stabilization of the rural population, an important issue in many developing countries (see Heal, 2004). In general in Southern Africa the alternative land use is cattle ranching, and tourism provides more and better-paid employment than ranching. The value of biodiversity in this role worldwide is hard to estimate, but must be in the hundreds of millions of dollars and possibly in the billions.

5.5 The Costs of Biodiversity Loss – Summary

In sum, biodiversity has great economic value although many aspects of this are hard to value. Its pharmacological value could be in the tens of billions of dollars. Intuitively, this order of magnitude makes sense: a successful pharmacological product can be worth USD 5 to USD 10 billion per year in revenues net of production costs, with a present value over its life of perhaps USD 50 to USD 100 billion. If there is the potential for several more blockbuster drugs in the biodiversity that remains, then this justifies the range of values that emerge from the Rausser-Small paper. That biodiversity's carbon storage value should be in the tens of billions of dollars also makes intuitive sense: forests are an integral part of biodiversity and store a large fraction of the planet's released carbon, a service for which there is now a market price on the European carbon markets. With carbon permits at roughly EUR 10 per ton, this service has immense value. If the world combats climate change effectively, this price will rise. The services of biodiversity in watersheds and in charismatic megafauna are harder to estimate in total, but again clearly run to billions of dollars.

In total therefore the loss of biodiversity could cost us certainly tens of billions of dollars, and possibly hundreds of billions, in the loss of the ecosystem services discussed above. It is important to recall that these are only some of the services provided by, and some of the sources of value of biodiversity. By way of example, the entire value of biodiversity as an input to agricultural progress is omitted, omitting its value in the production of hybrid corns amongst others. And the insurance value of biodiversity, clearly great from the example in section 1, is also omitted. These numbers are thus underestimates of the true value, perhaps serious underestimates. They are still high enough to make one take notice.

6. Conclusions

Ecosystems provide valuable services to humans. Biodiversity is an integral and essential part of most ecosystems and therefore of the provision of those services. There have been attempts at valuing some of the services provided by natural ecosystems, and these estimates provide some insight into the value of what we lose as we lose biodiversity, as we are doing. The estimates we have are underestimates, and in addition they neglect the option value calculations outlined in section 3. The true value of what we are losing is probably significantly larger, and as much of it will be gone before we decide to take action, we will never know the value of much of what we have lost.

We are uncertain about the value of biodiversity loss, although we know it is big enough to be serious. In the policy context, what matters is not so much the cost of policy inaction as a comparison of the costs of policies to stop biodiversity loss with the gains from such policies. So the operational question is: Could biodiversity conservation policies cost more than the value of the biodiversity that is being lost? Could it cost more than some hundreds of billions of dollars? There are almost certainly policies that would reduce biodiversity loss at a cost greatly below the value of the biodiversity thus conserved. As an example, consider the possibility of opening the European Union's Greenhouse Gas Emissions Trading Scheme to avoided deforestation carbon offsets, as has recently been discussed by the Union itself and a number of advisory bodies (Trines undated, Commission of the European Communities, 2005). As some of these references indicate, this could lead to the conservation of significantly more tropical forests and a mitigation of biodiversity loss on a large scale. There are as yet no systematic studies of the precise scale of forest conservation that would result from this, but it is generally thought to be large, and the cost to the EU would be negligible. So there clearly are policy options currently available that will conserve biodiversity at costs considerably less than the benefits from conservation.

A final observation about the value of biodiversity loss: there is considerable uncertainty about this value, not about the fact that it is large, but about exactly how large it is. If we as a society are risk-averse, then the possibility that we may be losing something even more valuable than we currently think is one that should concern us, particularly given the irreversible nature of the loss and the prospect of learning more about the values at stake in the future. Risk aversion, irreversibility and learning can justify a precautionary element to our behavior, which should tilt the balance in favor of conservation more than the numbers presented above do.

(Geoffrey Heal, Graduate School of Business and School of International and Public Affairs, Columbia University.)

Endnotes

1 Which is not to say that it is correct – for a discussion see National Academy of Sciences (2000).

2 There are also estimates of the value of wetlands as water purification systems. I have neglected these as the total extent of wetlands playing this role is small and the values are very context-dependent. These are discussed briefly in Pearce (2001).

3 Some have argued that the carbon sequestration and storage roles of tropical forests could be replaced by plantation forests, with much less biodiversity, suggesting that biodiversity is not needed to sequester and store carbon. That the carbon storage role of tropical rainforests could be played by monocultures of for example eucalypts is untrue: the subsoil communities of the tropical rainforests store more carbon than other ecosystems. The life of the trees is also more limited. It is also untrue that the carbon sequestration roles of tropical rainforests could be replaced by plantations, as the soil of tropical rainforests will generally not support plantation agriculture. Tropical rainforests have surprisingly little soil and what they have is rapidly lost on deforestation.

References

Armsworth P R, B E Kendall and F W Davis, "An Introduction to Biodiversity Concepts for Environmental Economists." *Resource and Energy Economics,* Volume 26, issue 2, June 2004 pp 115-136.

Arrow K J and L Hurwicz, "An Optimality Criterion for Decision-Making Under Ignorance." In Uncertainty and Expectations in Economics, Carter C F and J L Ford (eds) Basil Blackwell, Oxford, UK 1972.

Beissinger S R and N F R Snyder, *New World Parrots in Crisis: Solutions from Conservation Biology.* Smithsonian Institution Press, 1991.

Buttle Jonathan and Daniel Rondeau. "An Incremental Analysis of the Value of Expanding a Wilderness Area." *Canadian Journal of Economics,* Vol. 37, No. 1, pp. 189-198, February 2004.

Carpenter S R, D Ludwig and WA Brock, "Management and Eutroophication for Lakes Subject to Potentially Irreversible Change." *Ecological Applications* 1999, 9:751-771.

Chapin FS 3rd, *et al.*, "Consequences of Changing Biodiversity" Nature. 2000 May 11; 405(6783):234-42.

Commission of the European Communities. Communication from the Commission to the Council, the European Parliament, the European Economic and Social Committee and the Committee of the Regions: Winning the Battle against Global Climate Change. SEC (2005).

Conrad J M (2000), "Wilderness: options to preserve, extract, or develop". *Resource and Energy Economics,* 22.

Chetty R 2003, "A New Method of Estimating Risk Aversion." NBER Working Paper 9988 available at *www.nber.org/papers/w9988 Accessed June 15 2004.*

Daily G, (Editor) Nature's Services: Societal Dependence on Natural Ecosystems. Island Press 1997.

Dasgupta P S and G M Heal, Economic Theory and Exhaustible Resources. Cambridge University Press 1979.

Ecological Society of America. "Ecosystem Services: Benefits Supplied to Human Societies by Natural Ecosystems" Issues in Ecology, Number 2 Spring 1997.

Ehrlich P and J Roughgarden 1987. The Science of Ecology. Macmillan, New York.

Gollier C, *et al.* "Scientific Progress and Irreversibility: An Economic Interpretation of the Precautionary Principle." *Journal of Public Economics* 75:229-253, 2000.

Grime J P "Biodiversity and Ecosystem Function: the Debate Deepens." *Science* 277:1260 1997.

Heal G M Nature and the Marketplace. Island Press 2000.

Heal G M 2003 "Bundling Biodiversity" *Journal of the European Economic Association* April-May 2002 1(2-3): 553-560.

Heal GM, B Walker S, Levin K, Arrow P, Dasgupta G, Daily P, Ehrlich K-G, Maler N, Kautsky J, Lubchenco S, Schneider and D Starrett, "Genetic Diversity and Interdependent Crop Choice in Agriculture." Resource and Energy Economics 26 (2004): 175-184.

Holling C S "Resilience and Stability of Ecological Systems." *Annual Review of Ecological Systems* 4:1-23 1973.

Hooper D U and P M, Vitousek. "The Effect of Plant Composition and Diversity on Ecosystem Processes." *Science* 277: 1302-5. 1997.

Hughes J, *et al.* "Population Diversity: Its Extent and Extinction." *Science* 278:689-92. 1997.

Maskin E, "Decision Making under Ignorance with Implications for Social Choice Theory." *Theory and Decision* 11:319-37 1979.

McGrady-Steed J, *et al.* "Biodiversity Regulates Ecosystem Predictability." *Nature* 390: 162-5. 1997.

Millennium Ecosystem Assessment. "Ecosystems and Human Wellbeing." Available at *http://www.millenniumassessment.org/en/index.aspx*

Myers N "Biodiversity's Genetic Library" in Daily G (Editor) Nature's Services, Island Press 1997.

Naeem G P and S Li, "Biodiversity Enhances Ecosystem Reliability." *Nature* 390:505-9 1997.

National Academy of Sciences. Watershed Management for Potable Water Supply: Assessing the New York City Strategy. National Academies Press Washington D.C. 2000.

Nunes P A L D, J C J M Van Den Bergh and P Nijkamp, "Integration of Economic and Ecological Indicators of Biodiversity." In Valuation of Biodiversity Studies, OECD, Paris 2001.

OECD, "Economic Issues in Access and Benefit Sharing of Genetic Resources: a Framework for Analysis." Organization for Economic Cooperation and Development, November 2003, ENV/EPOC/GSP/BIO(2001)2/FIANL/

Pearce D, "Valuing Biological Diversity: Issues and Overview." In Valuation of Biodiversity Studies, OECD, Paris 2001.

Rausser G C and A A Small, "Valuing Research Leads: Bioprospecting and the Conservation of Genetic Resources." *Journal of Political Economy,* February 2000.

Simpson RD, R A, Sedjo and R Reid, "Valuing Biodiversity for Use in Pharmaceutical Research," *Quarterly Journal of Economics,* 1996 104 No.1:163-185.

Tilman D, *et al.* "Biodiversity and Ecosystem Properties." Science 278:1866-7. 1997.

Trines E, 'Possible role of Land Use, Land-Use Change and Forestry in future climate regimes: An inventory of some options, Report on the informal EU workshop on the future treatment of LULUCF." Study commissioned by the Netherlands Ministry of Agriculture, Nature and Food Quality from Eveline Trines, Marconistraat 27 2562 JB, Den Haag, the Netherlands.

Walters C J, *Adaptive Management of Renewable Resources.* McGraw Hill New York 1986.

5

The Structure of Biodiversity – Insights from Molecular Phylogeography

Godfrey M Hewitt

DNA techniques, analytical methods and palaeoclimatic studies are greatly advancing our knowledge of the global distribution of genetic diversity, and how it evolved. Such phylogeographic studies are reviewed from Arctic, Temperate and Tropical regions, seeking commonalities of cause in the resulting genetic patterns. The genetic diversity is differently patterned within and among regions and biomes, and is related to their histories of climatic changes. This has major implications for conservation science.

Introduction

Phylogeography, named by Avise *et al.,* in 1987 (1), is a recent and rapidly developing field that concerns the geographical distribution of genealogical lineages. It grew from the newly acquired technical ability to obtain DNA sequence variation from individuals across a species range, and from this to reconstruct phylogenies. These are then plotted geographically to display their spatial relationships and deduce the evolutionary origins and history of populations, subspecies and species (2,3). Genetic relationships between species

Source: Frontiers in Zoology, October 2004 © 2004, Hewitt. Reprinted with permission.

based on polytene chromosome banding patterns had been used earlier to deduce the geographic history of colonization and speciation by Drosophila of the Hawaiian Islands (4), and allozyme variation may still complement DNA information, but the ready access to mitochondrial (mt) DNA sequences opened the door to most animal species and generated this new field (5).

DNA Methods

Whilst mtDNA has lead the way in animal phylogeography, other DNA sequences are used, most commonly chloroplast (cp) in plants and non-coding nuclear (nc) regions in both animals and plants. MtDNA has a relatively fast rate of nucleotide divergence, well suited to examining events over the last few million years, but those of cpDNA and ncDNA are an order of magnitude lower and consequently less useful for such young divergences. More slowly evolving sequences are required for deeper phylogenetic history. For more recent events, like the last 10 thousand years, highly variable markers are needed, such as microsatellites and AFLPs, but whilst useful for population studies they suffer from homoplasy and produce equivocal genealogies (3).

Techniques for obtaining DNA sequence information are still advancing rapidly, with whole genome sequences being produced in a growing number of organisms. This allows sequences and markers to be identified and developed for many types of investigation, and some will be useful for phylogeographic studies. In particular, it is clear that genealogical data is required from several independent nuclear loci to provide a fuller and more reliable history of the species (6). Single nucleotide polymorphisms (SNPs) are also becoming available across the genome, which will produce comprehensive measures of genetic diversity and allow the construction of better population histories (7).

Analytical Approaches

The advance in DNA technology is producing a wealth of data for individuals, populations and species, and there are concomitant developments in analytical methods to define demographic history and evolutionary relationships, and to test their significance. This progress in analysis is facilitated by access to increasingly powerful desktop computers on which the increasingly sophisticated software can be used. Haplotype sequences of a particular DNA region can be ordered into a genealogical tree or network, and hence produce their phylogeny. When combined with their population frequency and geographic distribution, this provides a strong basis for inferences on the evolutionary history of the populations and species. The usual phylogeographic approach is to build a phylogeny from haplotype sequences using distance, parsimony and maximum likelihood methods and then represent the lineages geographically. There are several approaches that are regularly used, such as DNA Distance Phylogeography, Nested Clade Analysis, Haplotype Networks, Sequence Mismatch Distribution and Genetic and Demographic Simulation, e.g., PAUP and GeoDis (8-10). This last approach uses computers to explore broadly how DNA markers evolve in specified molecular, spatial and demographic conditions over history, and is being used increasingly (11). Recent developments seek to use the genetic data to estimate the demographic history of a population, the dates of historical bottlenecks or expansions, the size of ancestral populations, the location of refugial areas, the dates of divergence, the extent of migration and gene flow, the extent of fragmentation, and the sequence of such events to produce the present geographic distribution of genotypes, e.g. (12-15). We can expect further developments to provide even more discriminating analyses.

Each DNA sequence has its own genealogy and they may evolve at different rates. Furthermore, the various methods of analysis probe different aspects of the molecular and spatial history. Consequently, to reconstruct a species phylogeographic history one would ideally like to use a range of sequences (including nuclear, cytoplasmic, sex-linked, autosomal, conserved, neutral, high and low mutation rate) and apply a suite of pertinent analyses. This is not easy and often not possible with resources available. However, technological advances for molecules and computers have been explosive in the past decade, making much more detailed analysis possible today than only a few years ago, and this looks set to continue.

Paleoclimate and Paleobiology

The very different field of paleoclimatology is also experiencing great advances. The results of these are most pertinent to phylogeographic explanations, since they reveal the past environmental conditions and changes that have molded the evolutionary processes producing the present genetic structure. They provide a framework in which the phylogeny may be reconstructed. Past conditions can be deduced from carbon and oxygen isotopes, radiolarian skeletons, pollen grains and other residues from the sea bed, lake bottoms and ice sheets (16). Novel information sources like insect exoskeletons, coral terraces and stalagmites are adding to this (17-20). Such records show that earth's climate has been cooling for some 60 my with periodic (21, 41 and 100 kyr) global oscillations producing increasingly severe ice ages through the Quaternary (2.4 my). These involved greatly enlarged ice sheets and surrounding permafrost, and the lower temperatures and reduced water availability caused great changes in the distribution of species as demonstrated by the fossil record (16, 21). Nested within the major 100 kyr cycles are millennial scale oscillations, which can occur rapidly and are often severe (22, 23). Changes of 7–15°C may occur in decades and persist for centuries, as happened most recently in the Younger Dryas (11 kya), where fossils record shifts in the distributions of species.

These major changes in distributions of species occurred latitudinally as the ice sheets advanced and retreated, altitudinally in major mountain regions, and also longitudinally where new dispersal routes became available, as for example the Bering land bridge produced by the lowered sea level. The demographic fluctuations and adaptive challenges produced by such range changes would have had both stochastic and selective effects on the genetic variation and architecture, and the consequences of these can be studied by genetic and phylogeographic approaches. Thus the once distinct fields of paleobiology and phylogeography are now being combined, and have much to tell us about how present biodiversity was structured.

Fossil and Genetic Signals of Range Changes

With some effort it is now possible to obtain DNA data from specimens across the present range of a species, but fossil data is often more limited or absent. The most useful fossil data are for Europe and North America, which have extensive networks of pollen cores; a few span 3 ice ages (400 kyr), several reach the last interglacial (125 kyr), and a larger number cover back to the last glacial maximum (LGM 23-18 kyr) (21). There are also some helpful detailed series of beetle exoskeletons (24); animal bones and plant macrofossils tend to be localized and discontinuous, but are nonetheless useful markers of time and place. Reconstructions of paleovegetation have been made, e.g. (25), which are quite detailed from the LGM to the present, and when coupled with other fossil evidence indicate the extent and rapidity of changes in species distributions.

During the LGM the ice sheets and permafrost extended towards lower latitudes, so that generally species distributions were compressed toward the equator. Boreal species survived south of the ice in North America and Europe, but large areas of the north eastern Palearctic and Beringia remained ice free and some cold-hardy species appear to have survived here. Temperate species survived further south where habitats occurred to which they were each adapted. In Europe the disjunct southern peninsulas of Iberia, Italy and Balkans were particularly important, while in North America many temperate locations occurred around 40°N between the East and West coasts. Nearer the equator the pollen record is not extensive, but conditions were generally drier in the LGM and Tropical habitats were reduced while desert and savanna increased.

As a consequence, the habitats of many Boreal, Temperate and Tropical species were reduced and fragmented and they survived in refugia; but for some their habitats expanded, like those in the tundra and savanna. As the climate warmed after the LGM and the ice retreated, many Boreal and Temperate species were able to expand their ranges, as were some Tropical species. In some cases the refugial populations died out, but particularly in mountainous regions they could survive by ascending with the climate and their niche, as for example in the Alps, Andes, Appalachians and Arusha mountains. Such refugial regions allow the survival of species through several ice age cycles by ascending and descending to track their habitat, e.g. (26).

Such events modify the genetic content and structure of populations within species, and leave some traces for which we may search. Populations, races and subspecies that have been effectively separated for several glacial cycles will show divergence through the accumulation of neutral and possibly selected DNA changes. The extent of this divergence will be proportional to the time of separation. The haplotype tree or network of an evolving DNA sequence will reflect population expansions and contractions. Increasingly these effects can be analyzed, e.g. (27) and placed in some order of occurrence. When the geographic positions of haplotypes are included, a further range of deductions is possible.

For example, recently derived populations will contain a sample of the same haplotypes as the parent populations, which combined with paleo-information allows colonization routes to be deduced (28,29). The extent of distribution of younger haplotypes compared with that of older ones in the tree provides information on the past fragmentation of populations and processes involved in colonization (30); this can also be combined with paleo-information to deduce the intraspecific phylogeographic history.

Higher Latitudes – The Arctic

Most phylogeography has concerned Temperate biota (2,26), but recently a number of species from higher latitudes have been analysed in sufficient detail across their range to provide some first genetic insights into their biogeographic history. These include mammals, birds, fish, crustaceans and plants adapted to such cold conditions (31, 32). Table 1 contains some major studies of Holarctic animal species complexes. During the LGM the greatly extended Arctic ice sheets forced such species south, as evidenced by fossil records in Europe and North America. At the same time, large areas of Northeast Asia and the NW corner of North America were covered in permafrost but not glaciated. Fossil evidence suggests that these also contained refugia, particularly Beringia (18,33,34) which with lowered sea level joined Asia and America across the Bering Straits. The different range changes involved would be expected to have various effects on the genetic diversity that may have left marks of their occurrence and extent.

Distinct Parapatric Clades, Refugia and Range Changes

The phylogeographic structure, in terms of distinct regional DNA clades, is very marked in some species like the lemmings, voles and wren, moderate in the ptarmigan and dunlin, and less in the more mobile waterflea, reindeer and herring gull. The extent of DNA divergence between major clades in small mammals would suggest effective separation of up to 1 Myr, some 5-10 full glacial cycles, with further subdivision for shallower clades in more recent ice ages. In the gull, ptarmigan and reindeer the divergence among clades is low, indicating events occurring in the last or penultimate glacial cycles. Such recent structure would suggest that these species came from or were reduced to a small ancestral population in the late Pleistocene.

Table 1: Animal species with Holarctic ranges showing distinct phylogeographic pattern, with some indication of their possible divergence times, glacial refugia and genetic signals of population history. CA = Circumarctic, HA = High Arctic, PA = Palearctic, NA = Nearctic, BE = Beringia, GL = Greenland, NT = North Temperate.

Species Range	Phylogenetic Divergence (Myr)	Likely Refugia (fossil evidence *)	Genetic signals of range changes	Authors & Reference
Larus argentatus spp	9 clades	Atlantic	Allopatric fragmentation	Liebers *et al.* [96]
Herring gull complex	<0.5%	Aralo-Caspian	Bimodal mismatch	
CA HA	(0.1–0.3)		Recent expansions	
(+EurAsia Lakes)			Hybridizations	
Rangifer tarandus	7 clades	Beringia-Asia *	150 kyr expand	Gravlund *et al.* [97]
tundra reindeer	1–2%	W EurAsia*	15 kyr expand	Flagstad & Roed [98]
CA HA	(0.1–0.3)	N America	Ragged mismatch	
Lemmus ssp	4 clades	Beringia	Expand	Fedorov *et al.* [99]
true lemmings HA PA	3.8–7.9%	E Asia	Few haplotypes	
NA	(0.5–1.0)	Siberia*	Few haplotypes	
		N America*	Expand	
Dichrostonyx ssp	6 clades	Beringia*	Not to east	Fedorov & Stenseth [100]
collared lemmings HA	1–7%	Arctic Islands	Expand	
PA NA GL	(0.1–1.0)	E Asia	Low diversity	
		C&W Siberia*	Low diversity	
Microtus oeconomus	4 clades	Beringia	Not far	Brunhoff *et al.* [101]
root/tundra vole	2.0–3.5%	S Urals*	Few haplotypes	

Contd...

Contd...

HA PA BE	(0.2–0.6)	Caucasus	N expansion	
		C Europe*	N expansion	
Microtus agrestis	3 clades	S Urals*	Across Asia	Jaarola & Searle [102]
field vole	0.5–5.2%	Carpathians*	N&W expansion	
not HA, PA	(0.1–0.6)	Iberia	Not far	
Lagopus mutus	7 clades	Multiple, eg Greenland –	Several recent Expansions	Holder *et al.* [103, 104]
rock ptarmigan	0.21–1.12%	Beringia/Aleuts	Only 4% diversity within	
CA HA	(0.05–0.1)	Siberia	lineages	
Calidris alpina	5 clades	W Africa	W EurAsia	Wennerberg [105]
Dunlin (migrant)	1.1–3.3%	Arabia	Siberia	Wenink *et al.* [106]
CA	(0.1–0.3)	SE Asia	Beringia	
		C America	Canada	
Daphnia pulex	7 clades	Periglacial	Mixing, but some N Amer/	Weider *et al.* [107]
Waterflea (clonal)	0.5–3%	Some more local clones	Eurasian difference	
CA HA	(0.2–1.5)			
Troglodytes troglodytes	6 clades	NW Amer	Series of glacial vicariances	Drovetski *et al.* [38]
Winter wren	3.0–8.9%	NE Amer	Recent expansions	
NT not HA	(0.5–1.6)	NE Asia		
EurAsia BE N Amer		C Asia, S Europe		
Cleithrionomys rutilus/ glareolus/gapperi	12 clades	C Europe*	Several colonizations of N	Cook *et al.* [37]
red-backed voles	1–10%	E Asia*	Amer across Beringia from	
HA PA NA BE	(0.1–1.8)	Beringia	Asia	
		N Amer*		

The deeper clades of the true and collared lemmings, the root and field voles, and to some extent the shallow ones of the ptarmigan and dunlin, are remarkably parapatric and many contacts between them coincide around major features like the Urals, Lena, Kolyma and MacKenzie Rivers (Figure 1) (32). Regions where several subspecific and sister-specific boundaries coincide, called suture zones (35) have been recorded in North America and Europe, and are probably due to species having similar range changes and refugial areas (29). Thus these regional parapatric genomes seem to have been diverging separately over a number of ice ages, with distinct refugia from which they colonized to fill their individual interglacial distributions. The pattern of range changes may not be exactly the same through each cycle, but there is no sign of genetic mixing among mtDNA clades, and the major boundary features and refugia are likely to have been similar over the last few ice ages. Other taxon contacts occur in these regions in groups like birds and butterflies, and it will be informative to investigate their phylogeographies to look for commonalties and causes.

Each distinct regional clade would have had its own refugium, but the location of these remains to be determined. There are late glacial fossils of small mammals in Central Europe and also near the southern Urals, which were both parts of the extensive tundra and could have been refugia for European and West Asian clades. An additional refugium in east Asia would account for the clades meeting in the regions of the Lena and Kolyma Rivers, while fossils for a number of species in Beringia indicate that it was probably a significant refugium (32). A recent detailed examination of the phylogeography of the tundra vole in Beringia (36) pinpoints the contact between the Central Asian and Beringian clades on the Omolon River in the Kolyma uplands, which formed a partially glaciated barrier during the last ice age. This would obviously affect other species and provide the western boundary to many Beringian distributions. The equivalent boundary of Beringia in the east was formed by the North American ice sheets, which fringed Alaska from the MacKenzie River to the Aleutian Ranges.

The contractions, expansions and distant colonizations involved in these late Quaternary range changes would have influenced the genetic diversity, and should have left some signs. Low haplotype diversity and shallow clades are expected when populations have been severely contracted, and the age of the subsequent expansion may be gauged by mismatch analysis. The structure of haplotype networks and nested clades also provide indications of such events. Many of the populations and clades reported show these features (Table 1 and references). For example, tundra reindeer, herring gulls and rock ptarmigans have shallow clades diverging in the recent ice ages and populations with mismatch analyses indicating postglacial expansion and colonization of high Arctic regions. In those species like the lemmings and voles where DNA lineage divergence goes back several ice ages to perhaps the onset of more severe glaciations some 0.9 My, most clades are shallow with low diversity, particularly those for Central Asia, suggesting extensive colonization from a much reduced refugial source.

Figure 1

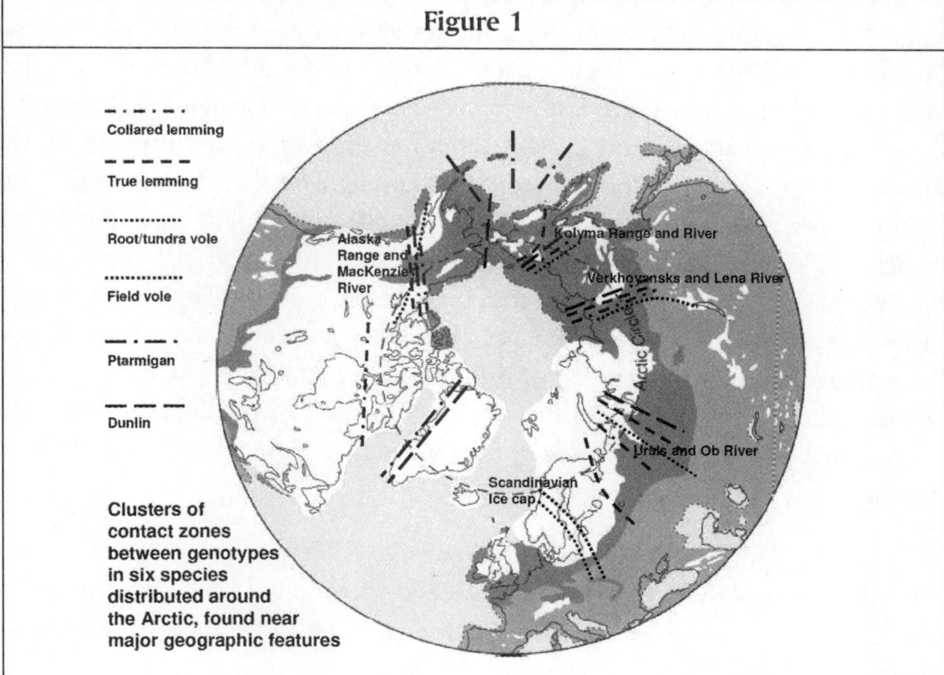

A polar projection showing the general regions of contact between diverged DNA clades of six Holarctic species (see text and A polar projection showing the general regions of contact between diverged DNA clades of six Holarctic species (see text and references for details and Latin binomials). Note the clustering near features like mountain ranges and major rivers. The Scandinavian cluster, which includes a number of other species, forms where the last remnants of the ice cap melted. Last glacial ice caps and sheets are in white, and tundra is darker grey. (I am grateful to Richard Abbott for the basic map).

This is an emerging feature of some significance. It does seem that even for cold adapted species life was hard in Central Asia during the ice age and it provided few refugia.

Beringia – An Arctic Refugium

Beringia, which spans Eastern Siberia and Alaska, was united and disjoined each Quaternary ice age by changes in the sea level of some 120m. It was only partly glaciated and fossil evidence shows it supported a mixture of tundra habitats, which allowed it to act as a causeway for continental migration between Asia and America for various species, including humans. The history of Beringia is currently of interest and some of the lower parts now under the Bering Sea probably had conditions suitable for mesic tundra species even in colder times (18). Several species have a distinct genetic clade across Beringia and some of these show higher diversity here than where previously glaciated regions were colonized, supporting its role as a glacial refugium (32). A careful analysis of genetic diversity in the tundra, or root vole, Microtus oeconomus through Beringia indicates that the Beringian voles probably expanded from a population with low genetic diversity, which had colonized from Central Asia in the penultimate glacial cycle (36). A broader phylogeographic study on related taxa of red-backed voles, Clethrionomys, demonstrated that North America has been colonized successfully from Asia at least twice, possibly three times in the mid-Pleistocene, and C. rutilus reached Alaska quite recently with the first Nearctic fossils in the Holocene (37). The phylogenies of a number of other species, including Homo sapiens also indicate Beringia as a colonization corridor.

Analysis of the DNA differences among these Clethrionomys species reveals that they have been diverging from the Early Pleistocene, whilst the divergence within the other listed species is much younger (Table 1). An exception is the winter wren, where there are six clades that appear to have begun diverging some 1.6 Mya, suggesting possible cryptic species (38). These distinct clades contain low diversity and signs of contraction and expansion. The species does not inhabit truly high Arctic habitats, reaching down to more Temperate latitudes. It would seem that its Holarctic range was progressively broken up by the increasingly severe Pleistocene glaciations, with connection across the oceans, Beringia and the centres of the continents becoming more difficult or impossible as its more Temperate range was forced south.

Mid Latitudes – Temperate Regions

So called Temperate species span a wide range of latitudes, since the climate determinants like insolation, oceans and altitude may produce suitable habitats between 20° and 60°. In Eurasia and North America those more cold hardy species generally have more northerly distributions than those better adapted to more southerly warmer climes, and this will also be reflected in altitudinal range differences. This would also be true of distributions in glacial periods, so that refugia for north Temperate species were nearer to the ice and permafrost than those of more southerly ones. Fossil evidence is very important in properly determining such Pleistocene range changes (39). In the Arctic species considered (Table 1), the winter wren and field vole provide examples of low Arctic/north Temperate species in what is a progression of adaptations, niches and ranges on the cold to warm axis from pole to equator.

Refugia and Colonization in Europe and North America

Paleoclimate and phylogeography have been most researched in Temperate Europe and North America, generating many particular examples and several more general conclusions (2, 26). In particular, the similarity among DNA haplotypes across a species range allows the deduction of which northern populations came from which southern populations, and their likely postglacial colonization routes. The fossil record underpins these deductions, particularly for refugial areas and times of colonization (26, 28). Thus many species survived the LGM in southern Europe, the centre and north being covered in tundra and ice. Fossil and DNA evidence show that the peninsulas of Iberia, Italy, Greece and the Balkans were major refugia and contributed variously to the postglacial recolonization of the north. Whilst a few mobile species crossed from North Africa, the Mediterranean Sea appears to have been a major barrier throughout the Quaternary. For many species these southern refugial areas currently contain much genetic diversity for haplotypes, lineages and subspecific taxa. These southern parts of Europe are also mountainous, so that species may survive by ascending and descending with climate changes. The extent of genetic divergence among these peninsular populations clearly indicates that for many species they have been effectively separate for several to many ice ages. Importantly, this separate refugial survival is seen as a causative factor in divergence and speciation (3, 29). It is possible to deduce pathways of genetic divergence in a geographical framework that has produced the continent's diversity of populations, subspecies and species.

The fossil record, in particular that of pollen and beetles, shows that postglacial colonization was to a large extent a property of the individual species niche and the distribution of its habitat. Nonetheless, the patterns of DNA divergence within species have some common features, which argue for common colonization processes and routes (29). Temperate species often show reduced haplotype diversity in the north, which is considered to be the result of rapid colonization with repeated founder events. This is seen in many species in Europe and North America, e.g. (28, 40-46). Furthermore the recolonized areas are often a broad

Figure 2

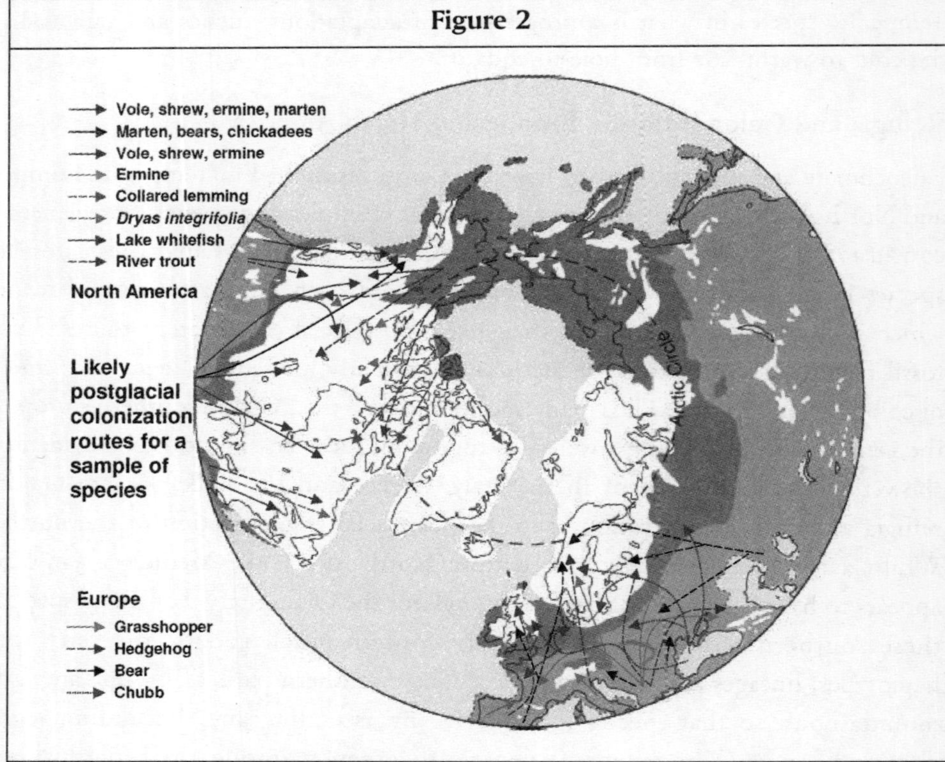

Likely postglacial colonization routes from refugial areas in Europe and North America for a distinctive sample of species that have been deduced from DNA haplotype relationships. Note that regions like central Scandinavia, Britain, the Pacific North West and central Canada contain a mixture of species whose genomes have come from different refugia (see text for discussion).

patchwork of distinct genomes that have emanated from the different refugia, and which usually form hybrid zones where they make contact. These hybrid zones in different species often appear to be clustered together and so may be considered to belong to suture zones (29, 35).

In Europe the Balkan haplotypes and genomes provided the main source for postglacial colonization for many species, while less came from Iberia and few came from Italy, probably hindered by the ice-capped Pyrenees and Alps. Species that exemplify these different patterns of colonization are the grasshopper, Chorthippus parallelus, the bear, Ursus arctos, and the hedgehog, Erinaceus europaeus/concolor (29). Freshwater fishes like the chubb, Leuciscus cephalus, often show colonization by different haplotypes up the Danube and Dneiper Rivers from the Black Sea (Figure 2) (32). Many European species phylogeographies are emerging and a considerable number broadly show these distribution patterns and probably followed similar colonization routes despite differences in their niche, mobility and life history. This apparent and remarkable commonality would seem to be a result of colonization following postglacial climate change in Europe's particular geography of southern peninsulas, transverse mountain ranges and northern plains. It demonstrates the explanatory power of combined phylogeography and paleoclimatology.

A particularly interesting consequence of these various colonization routes is that northern regions like Scandinavia and the Pacific NW of America have biotas that are mixtures of species whose genomes came from different refugia (26). Thus in Central Scandinavia the grasshopper genome came from the Balkans, the bear from Iberia and Russia, the hedgehog from Italy and the chubb from the Black Sea. For the NW corner of America, where Alaska, the Yukon and British Columbia meet, DNA evidence suggests that it has been colonized sequentially from four directions: (1) initially from the south along the coast by the long-tailed vole, dusky shrew, ermine and marten, (2) then from the south by an inland route by all but the marten, (3) from Beringia by the ermine, and (4) from the Appalachians by the marten, and possibly bears and chickadees (Figure 2). Several plant species also colonized probably along the two southern routes (32). Such mixed biotas carry a number of important implications. They mean that the component species from different refugia have not been evolving together during the previous glacial periods, so any close coadaptation must be either postglacial, or possibly survive from when their distant ancestors were sympatric. More general species-wide coadaptations may be maintained if the different species survive together in refugia. Where genomes from two or more refugia come together, genetic diversity will be increased by the presence of diverged lineages, as seen in hybrid and suture zones. Two populations in the same region living in similar habitats, but from different colonizing refugia will possess very different alleles and genomes, while conversely two very distant populations in distinct habitats may have the same refugial genome. This points to the importance of population history in the process of post glacial adaptation, and our understanding of it.

Mediterranean Latitudes – 30-40°N

The Mediterranean Sea, with Europe's refugial peninsulas and mountains in the north and North Africa to the south lies between roughly 30°N and 40°N, where at these latitudes in North America there are the Southern States across the Appalachians and Rockies to California. While in Europe the Scandinavian ice sheet came down to Warsaw about 52°N with extensive tundra to the south, the Laurentide ice sheet reached near 40°N. below the Great Lakes, with very little tundra and steppe. Such contrasts in geography have produced differences in the phylogeography of species, but there are also similarities. Species in these southern regions generally contain greater diversity for alleles, populations and subspecies, which form several distinct geographic genomic patches. The divergence among southern lineages and patches is often deeper than further north, indicating a longer survival and probably in the same region. This southern divergence is often estimated to be over many ice age cycles from the Early Pleistocene or even the Pliocene for some species (26). The best-studied regions in these latitudes are Iberia (47), South Eastern USA (2) and West Coast USA (48-51).

Molecular phylogeography began in the SE USA (1) and many terrestrial and aquatic species there have marked genetic substructure with concordant genomic boundaries (52). A number of recent studies in Iberia, covering a range of organisms, e.g., beetles (53), lizards (54), salamanders (55), woodmice (45), rotifers (47), and several plants, also show this intraspecific diversity and substructure, with lineages from as early as the Pliocene and often in mountain regions. Likewise West Coast phylogeography for salamanders (56), woodrats (51), shrews (50), frogs (57) and other species from California and the Cascades (49) reveals geographically structured lineages diverging from the Pliocene and Pleistocene. This pattern and divergence would seem to be a product of the geological and climate changes that have occurred, involving major mountain building and Quaternary ice ages. Lineages and populations from the northern parts of such southern regions, like the southern Appalachians, west and central Iberia and the northern Cascades, appear to have provided the main source for northward postglacial colonization, while genomes to the south survived with altitudinal shifts in broadly the same regions (32).

Species with distributions north and south of the Mediterranean Sea must have managed to cross this now major barrier to terrestrial organisms at some stage before or since the opening of the Gibraltar Straits some 5.3 Mya after the Messinian crisis (58). If this was before 5.3 Mya, they are likely to have diverged to sister species if there has been little gene flow. There are now phylogeographic studies on a few species both terrestrial and volant. In terrestrial species of salamanders Salamandra spp and scorpions Buthus spp (59,60) the DNA data shows that the N African/European divergence is old, before the opening of the Gibraltar Straits. While in the woodmouse Apodemus sylvaticus (45) the N African haplotypes appear recently derived from southern Iberia, possibly transferred by humans. Interestingly an old divergence in holm oak, Quercus ilex, may also perhaps be due to humans (61). Five flying species have been examined, of which the chaffinch, bearded vulture and barbastelle bat (62-64) have DNA phylogenies that indicate the Gibraltar Strait has not been a major barrier. In dragonflies Calopteryx spp (65) the North African genotypes are related to the Italian ones, and in honey bees (66) some African mtDNA haplotypes are found in south Iberia and Sicily, but nuclear markers indicate little migration. These latter two species and the bearded vulture appear to have crossed the Sicilian Channel, which was also narrow during the lowered sea level of the pleniglacials.

Lower Latitudes – Tropics and Savannah

Much of Africa, South America, South East Asia and North Australia lie in the Tropics. They are rich in species diversity, but with a few notable exceptions little is known of their phylogeography or their paleobiology (26). During the recent ice ages the climate was colder and drier in the Tropics, with increased deserts and

Table 2: Larger mammal species from across African savannah grasslands and woodlands showing phylogeographic pattern, with some indication of their phylogenetic structure and possible divergence times, refugia in west W, east E, and south S, and genetic signals of colonization and population history. LP = Late Pleistocene, MP = Mid Pleistocene, EP = Early Pleistocene, →= colonization/expansion. MS = Mismatch Expansion. All studies used d-loop mtDNA; also elephant used cytb and microsats, impala cytb, warthog and dog microsats, and buffalo Y.

Species	Phylogenetic structure	Likely Refugia (fossil evidence*)	Genetic signals of range changes	Reference
Hartebeest	3 step clades	W, E subdiv, S,	S→, E→,	Arctander *et al.* (83)
Alcelaphus buselaphus	W→S→E,	*0.7 My>	some shallow clades	
Topi	3 clades	(W), E, S,	S→, 2 clades, MS	Arctander *et al.* (83)
Damaliscus lunatus	S→E (+ S)	*0.7 My>	E→, shallow clade, MS	
Wildebeest	2 clades, S→E,	E, S,	S→E, MS in E	Arctander *et al.* (83)
Connochaetes taurinus	S structured	*1.5 My>	LP	
Kob (& Puku)	3 clades,	W, E, S	W E several times,	Birungi & Arctander (108)
Kobus kobus sl	W→E + S	* EP>	lineage mixing, MP LP	
Greater Kudu	3 clades,	E, S,	S→E LP	Nersting & Arctander (84)
Tragelaphus strepsiceros	E, SW→S→E	SW isolate	shallow clades	
		*widespread	diversity S>E	
Impala	2 clades,	(E), S	S→E LP	Nersting & Arctander (84)

Contd...

Contd...				
Aepyceros melampus	SW, S→E	SW isolate	network cluster E	
		*widespread	diversity S>E	
Wild dog	2 shallow clades,	(W), E, S	W→E&S?, 340 ky>	Girman *et al.* (89)
Lycaon pictus	E, S,		each clade 70 ky>	
	few haplotypes		mobile – mixing	
Buffalo	2 shallow clusters	W, E, S,	W→E <180 ky	Van Hooft *et al.* (85)
Syncerus caffer	W→E+S	* LP	E→S twice <130 ky	
	S nested in E		MS LP	
Elephant	3–5 clades	W, E, S,	W→S&E, EP	Nyakaana *et al.* (109)
Loxodonta africana	W→E+S→W	*EP	W→S&E, MP	Eggert *et al.* (88)
		*LP→	E→W, LP	
Warthog	3 distinct clades	W, E, S,	3 clades isolated MP by	Muwanika *et al.* (87)
Phacochoerus africanus	W→S→E	*0.78 My>	dry climate	
		*0.4 My→	contract/expand LP	

savannah and reduced rain forests. The pollen record is unfortunately poor, but it seems that forest species descended the mountains (~6°C lower LGM), while lowland forest species may have survived in many local wet places and gullies (16, 67, 68). It would seem clear that even tropical biotas have undergone repeated changes as a result of climatic oscillations through the Quaternary (26, 69).

Wet Tropics

Several phylogeographic studies from American and Australian rainforests and a few from Africa and Asia indicate that there is great genetic diversity produced by a complex history often diverging in the Pliocene (26, 70). A nice example of this has been studied in the montane forests of the central Divide of Costa Rica, where a North American salamander Bolitoglossa has radiated into tropical Middle America (71). There are strong allozyme and mtDNA differences between several populations only a few kilometres apart (Dnei 0.18, cytb 4%), and 2 putative species within 10 km (Dnei 0.45, cytb 9%). Such genetic distances indicate divergence from the late Pliocene through the Pleistocene. This has involved several adaptations to elevation zones that would have been amplified by local topographic isolation and climatic oscillations. These amphibians may have peculiar attributes, but phylogeographies of birds and freshwater fish in Middle America are also complex with many lineages (72). There are few studies yet, but it may be that the phylogeographic status of Bolitoglossa is not so unusual. Recently over 100 species of rhacophorine tree frog were described in Sri Lanka using mtDNA in combination with exophenotypic measures, when only 18 were previously known, and despite recent extinctions by Man's activities (73). This suggests that tropical biotas are not only amazingly diverse but highly structured genetically, both above and below the species level.

Genetic studies in tropical rainforests of SW Amazonia and NE Australia show phylogeographic divergence that is geographically concordant across a number of taxa and also originating in the Pliocene. The first set concern some 35 species of small mammals sampled along the Jurua River, and where for the majority there is a deep phylogenetic divide coincident with the Iquitos Arch. This formed as a bulge in front of the uplifting Andes in the Pliocene creating two basins that filled with sediment. The depth of mtDNA divergence between clades in these two basins places their separation at this time (74). The Iquitos Arch is also implicated in the phylogeographic structure of the dart-poison frog Epipedobates femoralis, which was also sampled along the Jurua River traversing this ancient ridge, and which also has coincident mtDNA divergence (cytb 12%) dating to the Pliocene (75). Interestingly, the collection sites differed markedly for their haplotypes, particularly in the headwaters region, again suggesting considerable local genetic structure. There has been a multitude of hypotheses proffered to explain the structure of Amazonian diversity, and such molecular phylogeographic approaches are beginning to distinguish amongst them.

The second set of phylogeographic studies involves several birds, reptiles and frogs from the remnant strip of tropical forest in NE Queensland (76). This wet forest has undergone contraction and fragmentation during the drier colder stages and expanded in the interglacials. These show concordant mtDNA divergence that possibly dates back to the Late Pliocene. This is coincident with the Black Mountain Corridor, a narrow region from which rainforest disappeared in the Pleistocene ice ages producing main north and south refugia. The north and south clusters of haplotypes have various structures, some of which show low diversity probably due to population contractions during the ice age, while others have retained more haplotype diversity possibly by survival in local patches of forest. The rainforest snail Gnarosophia bellend-enkerensis also shows these main phylogeographic features, and with some finer subdivisions. Its distribution from the LGM to the present has been modeled using current climate envelopes for this snail species mapped onto reconstructed paleoclimate distributions (77). There is good agreement between the changes in the modeled snail distribution from the LGM to the Holocene and the signals from mtDNA data of refugial locations and expansions. Such an approach provides support for the deductions of both paleoclimatic modeling and phylogeography. Furthermore, the relationship between particular paleoclimatic changes and genetic structure is sharpened by studies on other species from this region that are not adapted to rainforest, like grasshoppers and frogs (78, 79). These species show genetic subdivision that is coincident with other physical features that would provide refugia and barriers commensurate with their lifestyle and climatic history, such as the coastal humidity transition or Burdekin Gap.

The phylogeographic pattern demonstrated by amphibians, reptiles and small mammals, is also found in bird species from tropical Africa and South America, which show old Pliocene lineages in the lowland forest and mixed old and recently diverged clusters in the mountains. This has lead to the proposal that such mountains provided a relatively stable environment through the ice ages and rising mountains, in which older lineages survived and new ones were created (80). DNA divergence in spinetails from the Andes (81) and greenbuls from East Africa (82) provide evidence of montane speciation through the Pleistocene. Such mountain ranges appear to act as generators and reservoirs of lineages and species, and this probably is a function of their low latitude and topographic variety, which provide warm wet habitats through climatic and altitudinal range changes. The repeated small shifts in distribution driven by climate, along with continued uplifting of these mountains would provide conditions for contraction, selection, expansion and speciation.

Dry Tropics

There have been a number of recent DNA studies of several larger mammals from Africa that provide some interesting insights into how Pleistocene climatic changes modified their ranges and hence their genetic structure and divergence (Table 2). Whilst sampling such species across Africa is a major task, the threat posed by reductions in their numbers means that considerable efforts are being made to assess their genetic makeup for management and conservation, and many have a useful fossil record. They are not inhabitants of the wet forests, which were reduced during the colder drier glacial periods, but are found largely in the savannah grasslands and woodlands that had a different pattern of contraction and expansion. These habitats increased with the onset of the Pleistocene and its increasing glacial activity (3-2 Mya), with periods of dominance recorded around 1.7 and 1.2 Mya. They show increasing prevalence from 0.6 Mya through the Late Pleistocene, and fossils record the emergence of the associated mammal species and their subspecies since then (see references in Table 2). The DNA data, although not an accurate measure over this time scale, also indicates that these are recent events with most divergences in the last 0.4 My.

Most phylogeographies show some 3 major clades that are associated with 3 main areas of Tropical Africa, the west, east and south, indicating that these have been major refugial areas for the development of this divergence through climatic cycles in the Late Pleistocene (Figure 3). Many have shallow clades, mismatch analyses and star-like networks that are the expected result of contractions and expansions of these populations, and their phylogeographies indicate various colonizations between these major regions. In three species, the wildebeest Connochaetes taurinus, greater kudu Tragelaphus strepsiceros and the impala Aepyceros melampus, the data support colonization northwards to the east from refugia in the south of Africa (83, 84). The greater kudu and impala have distinct SW clades, which suggests isolation and survival there, as well as central South Africa, where many species appear to have had a refugium. Interestingly the first wildebeest fossils are from east Africa, so that the species seems to have disappeared from this region and been recolonized recently from the south.

Figure 3: Africa with Major Vegetation and Mountain Areas

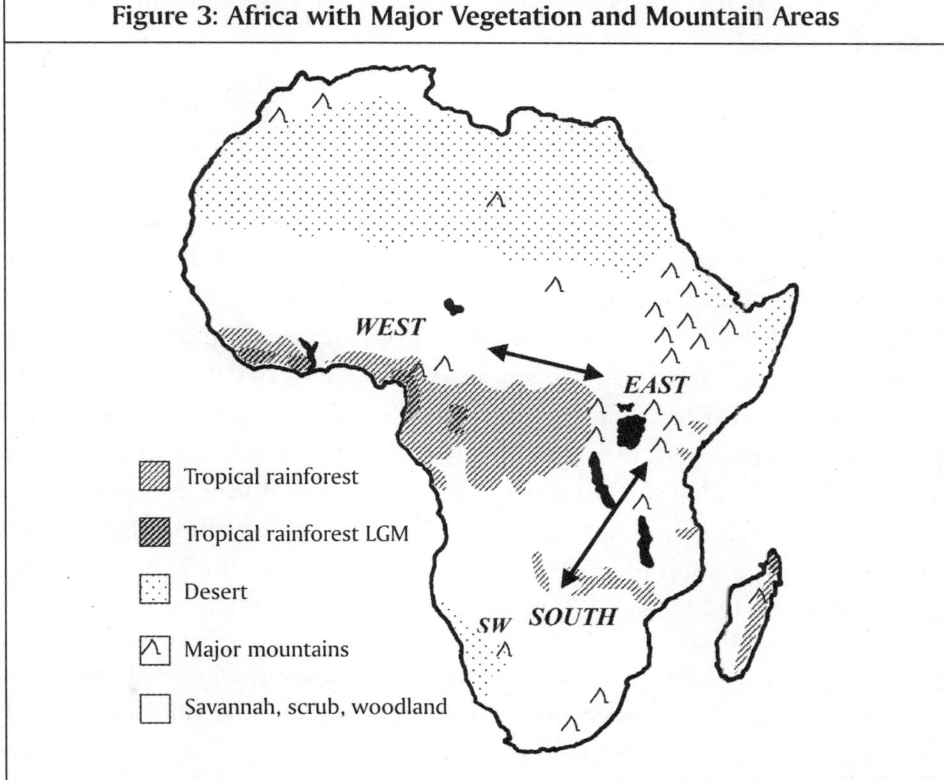

Africa with major vegetation and mountain areas. Reduced Tropical rainforest at the LGM is shown. The Savannah species often show West, East and South clades (see Table 2 for details) and the general areas of these are indicated. The genetic data also indicates colonisations between these possibly refugial areas in the middle and late Quaternary Period.

On the other hand, the Cape buffalo Synercus caffer caffer has younger haplotypes in the south, which along with mismatch analyses suggest one or two recent colonizations from populations in eastern Africa. The DNA divergence of these from central African buffalo subspecies is perhaps only 180-130 kyr, and roughly coincident with the Cape buffalo's genetic expansion and evidence from fossils (85). The hartebeest Alcelaphus buselaphus and the topi Damaliscus lunatus probably survived in a few places in southern and eastern Africa from which they expanded with better conditions (83, 86). The warthog Phacochoerus africanus is now relatively widespread, but has 3 distinct clades equivalent to subspecies, and with low divergence within each one (87). This points to strong isolation through the last few ice ages with considerable recent population reduction followed by population expansion.

The mtDNA phylogeny of the African elephant is more complexly structured, with haplotypes of the putative species of forest Loxodonta cyclotis and savannah L. africanus elephants mixed together in several clades (88). It suggests successive production of clades from the early Pleistocene, which involved colonization from the centre to the south and east with increasing savannah habitats, loss of competitors and punctuated by climatic cycles. Such admixture of clades in regions and taxa is probably a reflection of several such colonizations, and a recent invasion of western Africa from central regions is also indicated by the haplotype distribution. The African wild dog is very mobile and populations over the middle part of its current eastern through southern range show a mixture of haplo-types. There are 2 shallow but distinct mtDNA clades that would have diverged perhaps 340 kya, with each coalescing about 70 kya (89). The cause of this divergence is not clear, but restriction in habitat by climatic oscillations, and separate colonization of east and south from western Africa are possible.

Whilst there are common features within and distinguishing ones between the wet and dry examples reviewed from the Tropics, there are individualities to each species phylogeography; these reflect differences in biology and history that have produced differences in genetic structure. The richness and diversity seen at the species level is multiplied by that within species, and more studies are needed on biotas from such habitats to properly describe and understand them. Little is known from the species rich Tropics of South East Asia, or from the plains of South America and the large areas of Temperate Asia. There are a number of individual studies on a range of species, but one needs several representatives from each community to look for generalities. Given the species diversity in these areas, such genetic and phylogeographic knowledge is particularly important to inform sensible decisions on management and conservation of these resources.

Lessons for Conservation from Phylogeography

Man is in the unique position to know and predict the consequences for the environment and its biota of his innate will to survive and reproduce. Our actions are greatly modifying both of these, so we face the challenge of managing this sensibly. But we must also think of these natural and induced biotic changes in the light of future major changes in the climate. It is clear that global oscillations, producing great climatic changes, have occurred and will continue; in particular the increasingly severe Quaternary ice ages, which are now well researched and clearly demonstrated. Biotas have changed greatly due to these, and will do so again. We are currently well into an interglacial, the Holocene, and there is debate about how soon it will end and how quick this will be. If the north atlantic conveyor is turned off, and man may assist in this, colder conditions may return very quickly. On the other hand, in the shorter term global warming may continue, and man may assist this! What should we do about biodiversity, and how does phylogeography inform us for this?

The distribution of biodiversity across the world is largely measured as species diversity – their numbers, proportions and distinctness. But within a species there are often several geographic subspecies, and genetic studies have added greatly to knowledge of subspecific diversity, with some regions possessing more lineages and older divergence. Mountain ranges in warm Temperate and Tropical regions are seen to be important because they harbour much diversity at species, lineage and allelic levels. Phylogeographic studies reveal that this is likely a product of species surviving through climatic oscillations by tracking their habitat altitudinally and locally in a varied topography in regions not so affected by the extremes of climate change, e.g. (26). The southern mountains of Europe, the southern Appalachians and western mountains of USA have clearly been important as refugia and provided most colonists for the vast north temperate regions today. DNA divergence argues that this has happened repeatedly and so will probably happen again. To date there is little information on the patterns from Asia or South America, and it is needed. Such regions of Temperate refugial genetic diversity have accumulated lineages and alleles through several ice ages and deserve particular research and conservation.

The mountain regions in the wet Tropics of Africa and South America are very rich in species; while phylogeo-graphic studies reveal that these contain divergences often into the Pliocene with subsequent diversification through the Pleistocene. This retention of older diversity through millions of years along with younger lineages argues that they are both generators and reservoirs of divergence and species (26, 80). Less is known about the Tropical regions than Temperate ones, but the phylogeographic evidence does suggest that they can contain greater diversity and divergence in an area, as for example in the wet forests of Costa Rica or Queensland (71, 90). One wonders just what genetic diversity the species in the mountains of China and SE Asia contain. However, tropical Asia is even less studied than Africa and Central America, and with the rapid anthropogenic changes there studies on their phylogeography is urgent. With such information the genetic value of particular regions will be clearer, however their conservation and management involve complex and difficult political matters.

Besides these more general issues, there are a number of more particular lessons and questions. For example, it has recently been noted in several butterflies that the lower genetic diversity produced by postglacial colonization of northern Europe is correlated with their recent decline, as evidenced in national records. Moreover, different deduced postglacial colonization patterns show the same correlation of low genetic diversity and population decline (91). This suggests that the phylogeography of a species may be used as a predictor of demographic threat and loss. Clearly similar evidence from more species and groups is required to substantiate this.

Another particular example is the genetic diversity created by the formation of a hybrid zone as two genomes meet with postglacial colonization, which is multiplied when several species zones coincide as a suture zone, e.g. (26). It has been argued that such regions are important because of their genetic diversity (92). They may well allow the generation of occasional hybrid species (93, 94) and possibly reinforcement (95), but except perhaps for climatically very stable locations in the wet Tropics they are transient, and will disappear with each major climatic reversal. They may reform in roughly the same place each cycle, but fossil evidence suggests that this is not necessarily the case (29). Furthermore, for most zones the diversity they contain is accumulated in their refugia over several cycles, and hence these regions have greater long-term value.

The demonstration that the extent of divergence among lineages within and between sister species generally increases from the High Arctic to the wet Tropics reflects their evolutionary age. It can be argued that the richer Tropical biotas are more valuable than the poorer temperate ones, both in terms of their present allelic, lineage and species diversity, and their-long term survival. However, this overlooks the particular adaptations of temperate species and the vast highly productive biotas they produce. The agriculture of the temperate regions also supports much of the world's population. The consideration of regional diversity and adaptation raises a number of related questions. How well coadapted are recently assembled temperate ecosystems? Are north temperate and arctic species particularly selected for colonization by repeated range change? Are genetically richer genomes more able to adapt to change? Do putative refugial regions contain genetic variation that may be useful in agriculture? And there are many more such considerations.

Conclusion

The frequent major climatic oscillations in the last 2 My caused repeated changes in the ranges of surviving taxa, with extensive extinction and recolonization in higher latitudes and altitudinal shifts and complex refugia nearer the tropics. As a result of these past dynamics, the genetic diversity within species is highly structured spatially, with a patchwork of genomes divided by often coincident hybrid zones.

The extent of divergence among lineages within and between sister species generally increases from the High Arctic to the wet tropics and reflects their evolutionary age. Holarctic animal species show shallow but clear phylogeographic structure from the last or recent glaciations. Clades of several species are parapatric near major geographic features like rivers and mountains, suggesting they had similar range changes and refugial areas.

In temperate regions like Europe and North America there is much more diversity in the south, where it has accumulated in refugia over many ice ages, and much less in the north, where it was lost during postglacial colonization. These northern places have been colonized by species from different southern refugia, and have had little time to become closely coadapted. Furthermore, this loss of diversity in the north is implicated in the present reduction of population abundance in some species. Mammals from the Dry Tropics of Africa often show major clades in the west, east and south indicating major refugial areas for recent divergence through climatic cycles in the Late Pleistocene.

Mountain ranges in warm temperate and tropical regions would seem to be important for the survival of lineages through climatic changes, and hence for genome divergence and speciation. Such understanding of the distribution of biodiversity carries serious implications for the theory and practice of conservation.

(Godfrey M Hewitt, Honorary Professor, Chinese Academy of Science Beijing. He can be reached at g.hewitt@uea.ac.uk).

References

1. Avise JC, Arnold J, Ball RM, Bermingham E, Lamb T, Neigel JE, Reeb CA, Saunders NC: Intraspecific phylogeography: The mito-chondrial DNA bridge between population genetics and systematics. Annual Review of Ecology and Systematics 1987, 18:489-522.
2. Avise JC: Phylogeography: the history and formation of species. Cambridge, Mass., Harvard University Press; 2000:1-447.
3. Hewitt GM: Speciation, hybrid zones and phylogeography – or seeing genes in space and time. Molecular Ecology 2001, 10:537-549.
4. Carson HL: Chromosome tracers of the origin of species. Science 1970, 168:1414-1418.
5. Avise JC: The history and purview of phylogeography: A personal reflection. Molecular Ecology 1998, 7:371-379.
6. Zhang D-X, Hewitt GM: Nuclear DNA analyses in genetic studies of populations:practice, problems and prospects. Molecular Ecology 2003, 12:563-584.
7. Brumfield RT, Beerli P, Nickerson DA, Edwards SV: The utility of single nucleotide polymorphisms in inferences of population history. Trends in Ecology and Evolution 2003, 18:249-256.

8. Posada D, Crandall KA, Templeton AR: GeoDis: A program for the cladistic nested analysis of the geographical distribution of genetic haplotypes. Molecular Ecology 2000, 9:487-488.
9. Hewitt GM, Ibrahim KM: Inferring glacial refugia and historical migrations with molecular phylogenies. In Integrating Ecology and Evolution in a Spatial Context Edited by: Silvertown J and Antonovics J. Oxford, Blackwells; 2001:271-294.
10. Swofford DL: PAUP Phylogenetic Analysis Using Parsimony (and Other Methods), Version 4. Sunderland, Mass, Sinauer Associates.; 2001.
11. Ibrahim KM: Plague dynamics and population genetics of the desert locust: Can turnover during recession maintain population genetic structure? Molecular Ecology 2001, 10:581-591.
12. Beaumont MA: Recent developments in genetic data analysis: what can they tell us about human demographic history? Heredity 2004, 92:365-379.
13. Yoder AD, Yang Z: Divergence dates for Malagasy lemurs estimated from multiple gene loci: geological and evolutionary context. Molecular Ecology 2004, 13:757-773.
14. Nichols RA, Freeman KLM: Using molecular markers with high mutation rates to obtain estimates of relative population size and to distinguish the effects of gene flow and mutation: a demonstration using data from endemic Mauritian skinks. Molecular Ecology 2004, 13:775-787.
15. Templeton AR: Statistical phylogeography: methods of evaluating and minimizing inference errors. Molecular Ecology 2004, 13:789-809.
16. Williams D, Dunkerley D, DeDecker P, Kershaw P, J Chappell: Quaternary Environments. London, Arnold; 1998:1-329.
17. Yuan DX, Cheng H, Edwards RL, Dykoski CA, Kelly MJ, Zhang ML, Qing JM, Lin YS, Wang YJ, Wu JY, Dorale JA, An ZS, Cai YJ: Timing, duration, and transitions of the Last Interglacial Asian Monsoon. Science 2004, 304:575-578.
18. Elias SA, Berman D, Alfimov A: Late Pleistocene beetle faunas of Beringia: Where east met west. Journal of Biogeography 2000, 27:1349-1363.
19. Chappell J: Sea level changes forced ice breakouts in the Last Glacial cycle: New results from coral terraces. Quaternary Science Reviews 2002, 21:1229-1240.
20. Ponel P, Parchoux F, Andrieu-Ponel V, Juhasz I, de Beaulieu JL: A late-glacial-holocene fossil insect succession from Vallee des Merveilles, French Alps, and its paleoecological implications. Arctic Antarctic and Alpine Research 2001, 33:481-484.
21. Bennett KD: Evolution and Ecology: the Pace of Life. Cambridge, Cambridge University Press; 1997:1-241.

22. Holmgren CA, Penalba MC, Rylander KA, Betancourt JL: A 16,000 C-14 yr BP packrat midden series from the USA-Mexico Borderlands. Quaternary Research 2003, 60:319-329.

23. Hemming SR: Heinrich events: Massive late Pleistocene detritus layers of the North Atlantic and their global climate imprint. Reviews of Geophysics 2004, 42:RG1005.

24. Coope GR: The response of insect faunas to glacial-interglacial climatic fluctuations. Philosophical Transactions of the Royal Society of London Series B-Biological Sciences 1994, 344:19-26.

25. Prentice IC, Jolly D: Mid-Holocene and glacial-maximum vegetation geography of the northern continents and Africa. Journal of Biogeography 2000, 27:507-519.

26. Hewitt GM: The genetic legacy of the Quaternary ice ages. Nature 2000, 405:907-913.

27. Emerson BC, Paradis E, Thebaud C: Revealing the demographic histories of species using DNA sequences. Trends in Ecology and Evolution 2001, 16:707-716.

28. Hewitt GM: Some genetic consequences of ice ages, and their role in divergence and speciation. *Biological Journal of the Linnean Society* 1996, 58:247-276.

29. Hewitt GM: Post-glacial recolonization of European Biota. *Biological Journal of the Linnean Society* 1999, 68:87-112.

30. Templeton A: Nested clade analyses of phylogeographic data and testing hypotheses about gene flow and population history. Molecular Ecology 1998, 7:381-397.

31. Weider LJ, Hobaek A: Glacial refugia, haplotype distributions, and clonal richness of the Daphnia pulex complex in arctic Canada. Molecular Ecology 2003, 12:463-473.

32. Hewitt GM: Genetic consequences of climatic oscillations in the Quaternary. Philosophical Transactions of the Royal Society of London Series B-Biological Sciences 2004, 359:183-195.

33. Guthrie RD: Origin and causes of the mammoth steppe: a story of cloud cover, woolly mammal tooth pits, buckles, and inside-out Beringia. Quaternary Science Reviews 2001, 20:549-574.

34. Abbott RJ, Brochmann C: History and evolution of the arctic flora: in the footsteps of Eric Hultèn. Molecular Ecology 2003, 12:299-313.

35. Remington CL: Suture-zones of hybrid interaction between recently joined biotas. Evolutionary Biology 1968, 2:321-428.

36. Galbreath KE, Cook JA: Genetic consequences of Pleistocene glaciations for the tundra vole (Microtus oeconomus) in Beringia. Molecular Ecology 2004, 13:135-148.

37. Cook JA, Runck AM, Conroy CJ: Historical biogeography at the crossroads of the northern continents: molecular phylogenetics of red-backed voles (Rodentia : Arvicolinae). Molecular Phylogenetics and Evolution 2004, 30:767-777.

38. Drovetski SV, Zink RM, Rohwer S, Fadeev IV, Nesterov EV, Karagodin I, Koblik EA, Red'kin YA: Complex biogeographic history of a Holarctic passerine. Proceedings of the Royal Society of London Series B-Biological Sciences 2004, 271:545-551.

39. Tzedakis PC, Lawson IT, Frogley MR, Hewitt GM, Preece RC: Buffered tree population changes in a Quaternary refugium: Evolutionary implications. Science 2002, 297:2044-2047.

40. Soltis D, Gitzendanner M, Strenge D, Soltis P: Chloroplast DNA intraspecific phylogeography of plants from the Pacific Northwest of North America. Plant Systematics and Evolution 1997, 206:353-373.

41. Bernatchez L, C Wilson: Comparative phylogeography of nearctic and palearctic fishes. Molecular Ecology 1998, 7:431-452.

42. Austin JD, Lougheed SC, Neidrauer L, Chek AA, Boag PT: Cryptic lineages in a small frog: the post-glacial history of the spring peeper, Pseudacris crucifer (Anura: Hylidae). Molecular Phylo-genetics and Evolution 2002, 25:316-329.

43. Schmitt T, Giessl A, Seitz A: Postglacial colonization of western Central Europe by Polyommatus coridon (Poda 1761) (Lep-idoptera: Lycaenidae): evidence from population genetics. Heredity 2002, 88:26-34.

44. Trewick SA, Morgan-Richards M, Russell SJ, Henderson S, Rumsey FJ, Pinter I, Barrett JA, Gibby M, Vogel JC: Polyploidy, phylogeography and Pleistocene refugia of the rockfern Asplenium cete-rach: Evidence from chloroplast DNA. Molecular Ecology 2002, 11:2003-2012.

45. Michaux JR, Magnanou E, Paradis E, Nieberding C, Libois R: Mitochondrial phylogeography of the woodmouse (Apodemus sylvaticus) in the Western Palearctic region. Molecular Ecology 2003, 12:685-697.

46. Starkey DE, Shaffer HB, Burke RL, Forstner MRJ, Iverson JB, Janzen FJ, Rhodin AGJ, Ultsch GR: Molecular systematics, phylogeography, and the effects of Pleistocene glaciation in the painted turtle (Chrysemys picta) complex. Evolution 2003, 57:119-128.

47. Gomez A, Lunt DH: Refugia within refugia: patterns of phylogeographic concordance in the Iberian Peninsula. In Phylogeography of southern European refugia Edited by: Weiss S and Ferrand N Dordrecht, The Netherlands, Kluwer.; 2004.

48. Wake DB: Incipient species formation in salamanders of the Ensatina complex. Proceedings of the National Academy of Sciences USA 1997, 94:7761-7767.

49. Brunsfeld SJ, Sullivan J, Soltis DE, Soltis PS: Comparative phylogeography of north-western North America: A synthesis. In Integrating Ecology and Evolution in a Spatial Context Edited by: Silvertown J and Antonovics J, Oxford, Blackwells; 2001:319-339.

50. Maldonado JE, Vila C, Wayne RK: Tripartite genetic subdivisions in the ornate shrew (Sorex ornatus). Molecular Ecology 2001, 10:127-147.

51. Matocq MD: Phylogeographical structure and regional history of the dusky-footed woodrat, Neotoma fuscipes. Molecular Ecology 2002, 11:229-242.

52. Walker D, JC Avise: Principles of phylogeography as illustrated by freshwater and terrestrial turtles in the southeastern United States. Annual Review of Ecology and Systematics 1998, 29:23-58.

53. Gomez-Zurita J, Petitpierre E, Juan C: Nested cladistic analysis, phylogeography and speciation in the Timarcha goettingen-sis complex (Coleoptera, Chrysomelidae). Molecular Ecology 2000, 9:557-570.

54. Paulo OS, Jordan WC, Bruford MW, Nichols RA: Using nested clade analysis to assess the history of colonization and the persistence of populations of an Iberian Lizard. Molecular Ecology 2002, 11:809-819.

55. Alexandrino J, Arntzen JW, Ferrand N: Nested clade analysis and the genetic evidence for population expansion in the phylo-geography of the golden-striped salamander, Chioglossa lusi-tanica (Amphibia: Urodela). Heredity 2002, 88:66-74.

56. Moritz C, Schneider CJ, Wake DB: Evolutionary relationships within the Ensatina eschscholtzii complex confirm the ring species interpretation. Systematic Biology 1992, 41:273-291.

57. Nielson M, Lohman K, Sullivan J: Phylogeography of the tailed frog (Ascaphus truei): implications for the biogeography of the Pacific Northwest. Evolution 2001, 55:147-160.

58. Riding R, Braga JC, Martin JM, Sanchez-Almazo IM: Mediterranean Messinian Salinity Crisis: constraints from a coeval marginal basin, sorbas, southeastern Spain. Marine Geology 1998, 146:1-20.

59. Steinfartz S, Veith M, Tautz D: Mitochondrial sequence analysis of Salamandra taxa suggests old splits of major lineages and postglacial recolonizations of Central Europe from distinct source populations of Salamandra salamandra. Molecular Ecology 2000, 9:397-410.

60. Gantenbein B, Largiader CR: The phylogeographic importance of the Strait of Gibraltar as a gene flow barrier in terrestrial arthropods: A case study with the scorpion Buthus occitanus as model organism. Molecular Phylogenetics and Evolution 2003, 28:119-130.

61. Lumaret R, Mir C, Michaud H, Raynal V: Phylogeographical variation of chloroplast DNA in holm oak (Quercus ilex L.). Molecular Ecology 2002, 11:2327-2336.

62. Griswold CK, Baker AJ: Time to the most recent common ancestor and divergence times of populations of common chaffinches (Fringilla coelebs) in Europe and North Africa: Insights into pleistocene refugia and current levels of migration. Evolution 2002, 56:143-153.

63. Godoy JA, Negro JJ, Hiraldo F, Donazar JA: Phylogeography, genetic structure and diversity in the endangered bearded vulture (Gypaetus barbatus, L.) as revealed by mitochondrial DNA. Molecular Ecology 2004, 13:371-390.

64. Juste J, Ibanez C, Trujillo D, Munoz J, Ruedi M: Phylogeography of Barbastelle bats (Barbastella barbastellus) in the western Mediterranean and the Canary Islands. Acta Chiropterol 2003, 5: 165-175.

65. Weekers PHH, De Jonckheere JF, Dumont HJ: Phylogenetic relationships inferred from ribosomal ITS sequences and biogeographic patterns in representatives of the genus Calopteryx (Insecta : Odonata) of the West Mediterranean and adjacent West European zone. Molecular Phylogenetics and Evolution 2001, 20:89-99.

66. Franck P, Garnery L, Loiseau A, Oldroyd BP, Hepburn HR, Solignac M, Cornuet J-M: Genetic diversity of the honeybee in Africa: Microsatellite and mitochondrial data. Heredity 2001, 86:420-430.

67. Colinvaux PA, DeOliveira PE, Moreno JE, Miller MC, Bush MB: A long pollen record from lowland Amazonia: Forest and cooling in glacial times. Science 1996, 274:85-88.

68. Flenley JR: Tropical forests under the climates of the last 30,000 years. Climatic Change 1998, 39:177-197.

69. Willis KJ, Gillson L, Brncic TM: How "virgin" is virgin rainforest? Science 2004, 304:402-403.

70. Moritz C, Patton JL, Schneider CJ, Smith TB: Diversification of Rainforest Faunas: An integrated molecular approach. Annual Reviews of Ecology and Systematics 2000, 31:533-563.

71. Garcia-Paris M, Good DA, Parra-Olea G, Wake DB: Biodiversity of Costa Rican salamanders: Implications of high levels of genetic differentiation and phylogeographic structure for species formation. Proceedings of the National Academy of Sciences USA 2000, 97:1640-1647.

72. Bermingham E, Martin AP: Comparative mtDNA phylogeography of neotropical freshwater fishes: Testing shared history to infer the evolutionary landscape of lower Central America. Molecular Ecology 1998, 7:499-517.

73. Meegaskumbura M, Bossuyt F, Pethiyagoda R, Manamendra-Arachchi K, Bahir M, Milinkovitch MC, Schneider CJ: Sri Lanka: An Amphibian Hot Spot. Science 2002, 298:379-379.

74. Patton JL, da Silva MNF: Rivers, Refuges, and Ridges: The Geography of Speciation of Amazonian Mammals. In Endless Forms: Species and Speciation Edited by: Howard D and Berlocher S. Oxford, Oxford Univ. Press; 1997:202-213.

75. Lougheed SC, Gascon C, Jones DA, Bogart JP, Boag PT: Ridges and rivers: A test of competing hypotheses of Amazonian diversification using a dart-poison frog (Epipedobates femoralis). Proceedings of the Royal Society of London Series B-Biological Sciences 1999, 266:1829-1835.

76. Schneider C, Cunningham M, Moritz C: Comparative phylogeography and the history of endemic vertebrates in the Wet Tropics rainforests of Australia. Molecular Ecology 1998, 7: 487-498.

77. Hugall A, Moritz C, Moussalli A, Stanisic J: Reconciling paleodistri-bution models and comparative phylogeography in the Wet Tropics rainforest land snail Gnarosophia bellendenkerensis (Brazier 1875). Proceedings of the National Academy of Sciences USA 2002, 99:6112-6117.

78. Shaw DD, Marchant AD, Contreras N, Arnold ML, Groeters F, Kohl-mann BC: Genomic and Environmental Determinants of a Narrow Hybrid Zone: Cause or Coincidence? In Hybrid zones and Evolutionary Process Edited by: Harrison R G. Oxford, Oxford Univ. Press; 1993:165-195.

79. Schauble CS, Moritz C: Comparative phylogeography of two open forest frogs from eastern Australia. *Biological Journal of the Linnean Society* 2001, 74:157-170.

80. Fjeldsa J, Lovett JC: Geographical patterns of old and young species in African forest biota: The significance of specific montane areas as evolutionary centres. Biodiversity and Conservation 1997, 6:323-244.

81. Garcia-Moreno J, Arctander P, Fjeldsa J: A case of rapid diversification in the Neotropics: Phylogenetic relationships among Cranioleuca spinetails (Aves, Furnariidae). Molecular Phylogenetics and Evolution 1999, 12:273-281.

82. Roy M: Recent diversification in African greenbuls (Pycnon-otidae: Andropadus) supports a montane speciation model. Proceedings of the Royal Society of London Series B-Biological Sciences 1997, 264:1337-1344.

83. Arctander P, Johansen C, Coutellec-Vreto M: Phylogeography of three closely related African bovids (tribe Alcelaphini). Molecular Biology and Evolution 1999, 16:1724-1739.

84. Nersting LG, Arctander P: Phylogeography and conservation of impala and greater kudu. Molecular Ecology 2001, 10:711-719.

85. Van Hooft WF, Groen AF, Prins HT: Phylogeography of the African buffalo based on mitochondrial and Y-chromosomal loci: Pleistocene origin and population expansion of the Cape buffalo subspecies. Molecular Ecology 2002, 11:267-279.

86. Flagstad O, Syvertsen PO, Stenseth NC, Jakobsen KS: Environmental change and rates of evolution: the phylogeographic pattern within the hartebeest complex as related to climatic variation. Proceedings of the Royal Society of London Series B-Biological Sciences 2001, 268:667-677.

87. Muwanika VB, Nyakaana S, Siegismund HR, Arctander P: Phylogeography and population structure of the common warthog (Phacochoerus africanus) inferred from variation in mitochondrial DNA sequences and microsatellite loci. Heredity 2003, 91:361-372.

88. Eggert LS, Rasner CA, Woodruff DS: The evolution and phylogeography of the African elephant inferred from mitochondrial DNA sequence and nuclear microsatellite markers. Proceedings of the Royal Society of London Series B-Biological Sciences 2002, 269:1993-2006.

89. Girman DJ, Vila C, E Geffen, Creel S, Mills MGL, McNutt JW, Ginsberg J, Kat PW, Mimiya KH, Wayne RK: Patterns of population subdivision, gene flow and genetic variability in the African wild dog (Lycaon pictus). Molecular Ecology 2001, 10:1703-1723.

90. Moritz C: Strategies to protect biological diversity and the evolutionary processes that sustain it. Systematic Biology 2002, 51:238-254.

91. Schmitt T, Hewitt GM: The genetic pattern of population threat and loss: a case study of butterflies. Molecular Ecology 2004, 13:21-31.

92. Smith TB, Wayne RK, Girman DJ, Bruford MW: A role for eco-tones in generating rainforest biodiversity. Science 1997, 276:1855-1857.

93. Arnold ML: Natural Hybridization and Evolution. Oxford, Oxford University Press; 1997:1-215.

94. Rieseberg LH, Raymond O, Rosenthal DM, Lai Z, Livingstone K, Nakazato T, Durphy JL, Schwarzbach AE, Donovan LA, Lexer C: Major ecological transitions in wild sunflowers facilitated by hybridization. Science 2003, 301:1211-1216.

95. Marshall JL, Arnold ML, Howard DJ: Reinforcement: The road not taken. Trends in Ecology and Evolution 2002, 17:558-563.

96. Liebers D, De Knijff P, Helbig AJ: The herring gull complex is not a ring species. Proceedings of the Royal Society of London Series B-Biological Sciences 2004, 271:893-901.

97. Gravlund P, Meldgaard M, Paabo S, Arctander P: Polyphyletic origin of the small-bodied, high-arctic subspecies of tundra reindeer (Rangifer tarandus). Molecular Phylogenetics and Evolution 1998, 10:151-159.

98. Flagstad O, Roed KH: Refugial origins of reindeer (Rangifer tarandus L.) inferred from mitochondrial DNA sequences. Evolution 2003, 57:658-670.

99. Fedorov VB, Goropashnaya AV, Jaarola M, Cook JA: Phylogeogra-phy of lemmings (Lemmus): No evidence for postglacial colonization of Arctic from the Beringian refugium. Molecular Ecology 2003, 12:725-731.

100. Fedorov VB, Stenseth NC: Multiple glacial refugia in North American Arctic: inference from phylogeography of the collared lemming (Dicrostonyx groenlandicus). Proceedings of the Royal Society of London Series B-Biological Sciences 2002, 269:2071-2077.

101. Brunhoff C, Galbreath KE, Fedorov VB, Cook JA, Jaarola M: Holarctic phylogeography of the root vole (Microtus oeconomus): implications for late Quaternary biogeography of high latitudes. Molecular Ecology 2003, 12:957-968.

102. Jaarola M, Searle JB: Phylogeography of field voles (Microtus agrestis) in Eurasia inferred from mitochondrial DNA sequences. Molecular Ecology 2002, 11:2613-2621.

103. Holder K, Montgomerie R, Friesen VL: A test of the glacial refugium hypothesis using patterns of mitochondrial and nuclear DNA sequence variation in rock ptarmigan (Lagopus mutus). Evolution 1999, 53:1936-1950.

104. Holder K, Montgomerie R, Friesen VL: Glacial vicariance and historical biogeography of rock ptarmigan (Lagopus mutus) in the Bering region. Molecular Ecology 2000, 9:1265-1278.

105. Wennerberg L: Breeding origin and migration pattern of dunlin (Calidris alpina) revealed by mitochondrial DNA analysis. Molecular Ecology 2001, 10:1111-1120.

106. Wenink PW, Baker AJ, Rosner H-U, Tilanus MGJ: Global mito-chondrial DNA phylogeography of Holarctic breeding dun-lins (Calidris alpina). Evolution 1996, 50:318-330.

107. Weider LJ, Hobaek A, Colbourne JK, Crease TJ, Dufresne F, Hebert PDN: Holarctic phylogeography of an asexual species complex I. Mitochondrial DNA variation in Arctic Daphnia. Evolution 1999, 53:777-792.

108. Birungi J, Arctander P: Large sequence divergence of mitochon-drial DNA genotypes of the control region within populations of the African antelope, kob (Kobus kob). Molecular Ecology 2000, 9:1997-2008.

109. Nyakaana S, Arctander P, Siegismund HR: Population structure of the African savannah elephant inferred from mitochondrial control region sequences and nuclear microsatellite loci. Heredity 2002, 89:90-98.

6

Design and Applicability of DNA Arrays and DNA Barcodes in Biodiversity Monitoring

Mehrdad Hajibabaei, Gregory A C Singer, Elizabeth L Clare and Paul D N Hebert

Background: The rapid and accurate identification of species is a critical component of large-scale biodiversity monitoring programs. DNA arrays (micro and macro) and DNA barcodes are two molecular approaches that have recently garnered much attention. Here, we compare these two platforms for identification of an important group, the mammals.

Results: Our analyses, based on the two commonly used mitochondrial genes cytochrome c oxidase I (the standard DNA barcode for animal species) and cytochrome b (a common species-level marker), suggest that both arrays and barcodes are capable of discriminating mammalian species with high accuracy. We used three different datasets of mammalian species, comprising different sampling strategies. For DNA arrays we designed three probes for each species to address intraspecific variation. As for DNA barcoding, our analyses show that both cytochrome c oxidase I and cytochrome b genes, and even smaller fragments of

Source: BMC Biology, June 2007.

(mini-barcodes) can successfully discriminate species in a wide variety of specimens.

Conclusion: This study showed that DNA arrays and DNA barcodes are valuable molecular methods for biodiversity monitoring programs. Both approaches were capable of discriminating among mammalian species in our test assemblages. However, because designing DNA arrays require advance knowledge of target sequences, the use of this approach could be limited in large-scale monitoring programs where unknown haplotypes might be encountered. DNA barcodes, by contrast, are sequencing-based and therefore could provide more flexibility in large-scale studies.

Background

Species identification is essential for large-scale biodiversity monitoring and conservation (1). Several molecular methods have been employed for biodiversity studies, but traditional methods such as allozyme analysis are usually laborintensive and irreproducible. Because of advances in DNA-based technologies, approaches such as DNA arrays and DNA barcoding have recently gained attention. Both of these methods are based on comparative DNA sequence analysis, but they have significant differences.

Micro- and macro-arrays rely on the hybridization of short (i.e., 25 base) specific nucleotide probes to DNA from the target organism and subsequent detection of the hybridization signal. Although array-based technologies have been widely used in gene expression studies, their use in biodiversity research has been less rigorous, mainly targeting pathogenic microorganisms (2) and arrays of environmental samples (3). Pfunder *et al.* (4), however, have advocated an array-based method for the identification of voles and shrews for biodiversity monitoring. Although this study focuses on a limited number of species, the authors have predicted that such an approach can be used for the development of a so called 'Mammalia Chip', in the case of mammalian species, or even a 'Biodiversity Chip' for monitoring key species of different taxa from bacteria to mammals (4).

Species identification by DNA barcoding is based on sequencing a short standardized genomic region of the target specimen and comparing this information to a sequence library from known species (5). The proposed standard barcode sequence for animal species is a 650-bp fragment of the mitochondrial gene cytochrome c oxidase I (COI, cox1). This DNA barcode has successfully been used for the identification of species in various vertebrate and invertebrate groups from birds to Lepidoptera (6-8), and in different geographical settings from the arctic to the tropics (6, 9). Additionally, smaller fragments (i.e., 100 bases) of the standard COI barcode – 'mini-barcodes' – have been shown to be effective for species identification in specimens whose DNA is degraded or potentially in other situations where obtaining a full-length barcode is not feasible (10). Barcoding is now being extended to other groups such as fungi, plants and protists, and the Barcode of Life Initiative has gained international momentum by the establishment of the Consortium for the Barcode of Life (CBOL), which plans to assemble DNA barcode libraries for all fish and birds (11).

Here, we compare the design and applicability of both array-based and barcoding platforms for specimen identification in mammalian species. We have chosen mammals because they constitute an important target for biodiversity studies and include many endangered species. However, mammalian species have not been broadly targeted for developing array-based or barcoding identification systems previously. A rapid identification method will aid in the tracking of illegal trafficking of mammalian species and their tissues. We have selected two mitochon-drial loci for our analysis: COI – the proposed standard animal DNA barcode – and cytochrome b (cytb), which is commonly used as a species-level marker and particularly so in mammalian biosystematics (4, 12). We used both of these genes to test the possibility of designing a Mammalia Chip. We also used these sequences and various size fragments of them to test the feasibility of DNA barcoding analysis for mammalian species. We targeted three data-sets of mammalian species for these analyses: 121 species across the taxonomy of mammals (mammalian dataset), a dense sampling of 87 species of neotropical bats (bat dataset), and a wide geographical sampling of a single genetically diverse bat species (Sturnira lilium dataset).

Results

Array-Based Analysis

Both COI and cytb performed well as templates for probe design in mammalian dataset. However, we observed a sharp decline in the number of unique probes in species as haplotype diversity increased (Figure 1). For example, no 25mer probe in either COI or cytb was shared by all humans but also distinct from other species. We dealt with this limitation by choosing three probes from each species so that at least two probes should exactly match sequences within the target species. Another consideration was to ensure that the probes match among the first 150 bases from the 5' end of the target genes. This is important as COI and cytb are longer than 1 KB and are difficult to amplify in their entirety in samples with degraded DNA (i.e., traces of tissues, processed material and archival specimens) (10). Our algorithm provided COI and cytb probes for 90.9% and 98.4% of the species in this mammalian dataset, respectively (Additional file 1). As for the bat dataset, we found a somewhat similar result (using available COI sequences) and were able to design probes for 89.7% of the species in this assemblage (Additional file 1).

DNA Barcoding Analysis

Whole COI and cytb delivered similar results for the mammalian dataset, identifying all the species in our assemblage in a Neighbor-Joining (NJ) analysis (13) (Table 1). The standard animal barcode – a 650 bp fragment at the 5' end of COI – identified 96.7% of the species (Table 1). The same fragment size of cytb provided 98.3% species-level resolution (Table 1). Significantly, mini-barcodes of COI and cytb were also capable of discriminating among species of mammals, although the resolution was somewhat lower (Table 1). Interestingly, in the COI data we found that a mini-barcode positioned at nucleotides 437-654 (mini-barcode 5 in Table 1) provided the same resolution for species identification as the standard barcode sequence. In contrast, all the cytb mini-barcodes provided lower resolution as compared to a barcode-size fragment of the cytb gene (Table 1). The results obtained in the NJ analysis were confirmed when we plotted the sequence length against the probability of obtaining a unique sequence for each species. Interestingly, we found that the minimum signal required to provide unique barcodes in about 95% of the species in the mammalian dataset is a short ~50 base fragment of the 5' region of either the COI barcode or cytb gene, but the resolution decreases sharply with smaller sequences (Figure 2).

Figure 1: Microarray Probes

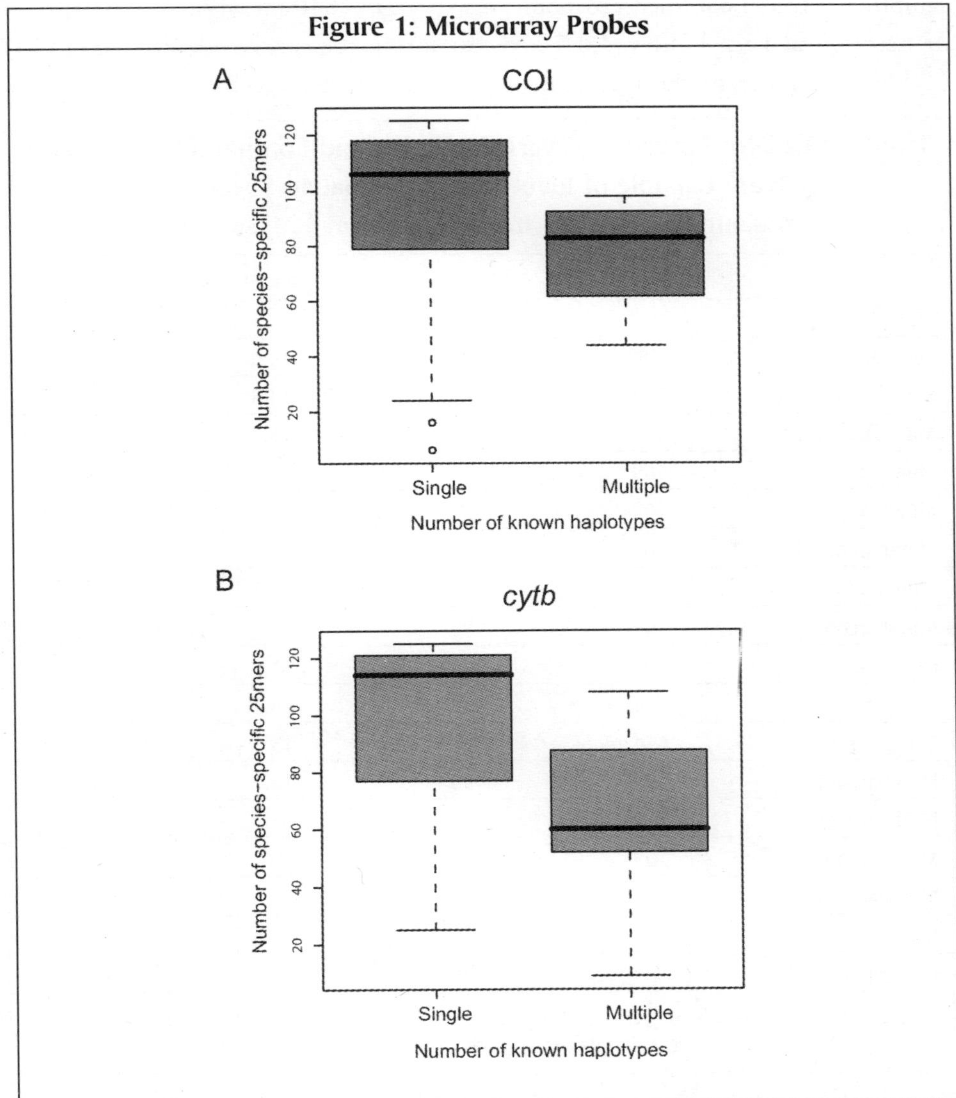

Species with only one representative sequence were easy to design probes for. However, it is more challenging to find probes that were unique within species but capable of distinguishing between species when that species has several known haplotypes. Data is from 150 bases of the 5' region of COI (A) and *cytb* (B).

An evaluation of COI barcodes in a dataset with lower taxonomic diversity (compared to our mammalian dataset) but with a somewhat higher density of sampling within a confined taxonomic assemblage – 840 individuals of 87 species of neotropical bats – showed a 100% resolution for species identification (14) (Table 2). Similar to the mammalian dataset, mini-barcodes of 109 bases were

Table 1: COI DNA Barcodes of Varied Lengths and Comparable Fragments of Cytb are Capable of Identifying Mammalian Species in Test Assemblage of 1585 Individuals from 121 Species.

	Length	% Res	% K2P	% Var
COI				
Full gene	1557	100	14.9	56.3
Standard barcode	654	96.7	14.8	55.8
Mini-barcode 1	109	93.3	19.7	61.5
Mini-barcode 2	109	95	11.7	51.4
Mini-barcode 3	109	93.3	15.8	57.8
Mini-barcode 4	109	95	16.7	56.0
Mini-barcode 5	109	96.7	13.5	54.1
Mini-barcode 6	109	95	12.6	54.1
cytb				
Full gene	1149	100	16.9	69.3
Barcode size	654	98.3	15.9	65.9
Mini-barcode 1	109	95	15.7	67.9
Mini-barcode 2	109	93.3	17.3	65.1
Mini-barcode 3	109	96.7	16	68.8
Mini-barcode 4	109	95	13.8	58.7
Mini-barcode 5	109	95	15.3	65.1
Mini-barcode 6	109	95	18.2	69.7

Res, resolution in neighbor-joining analysis (13); K2P, genetic distances based on Kimura two-parameter nucleotide substitution model (20); var, variable sites.

also capable of discriminating among more than 95% of the species in this bat dataset (Table 2). In addition, the minimum signal required to provide unique barcodes in more than 95% of the species of bats was a short ~30 base fragment of the 5' region of COI (results not shown). Comparison of COI and cytb in 34 individuals of one of these species, Sturnira lilium, across 13 sampling localities in Central and South America suggests that both genes provide similar resolution and can detect three geographical variants within this species (Figure 3). Similar resolution is achieved by using mini-barcodes of both COI and cytb for this species (results not shown).

Figure 2: Surprisingly Short Barcode Sequences are Capable of Distinguishing Between Species

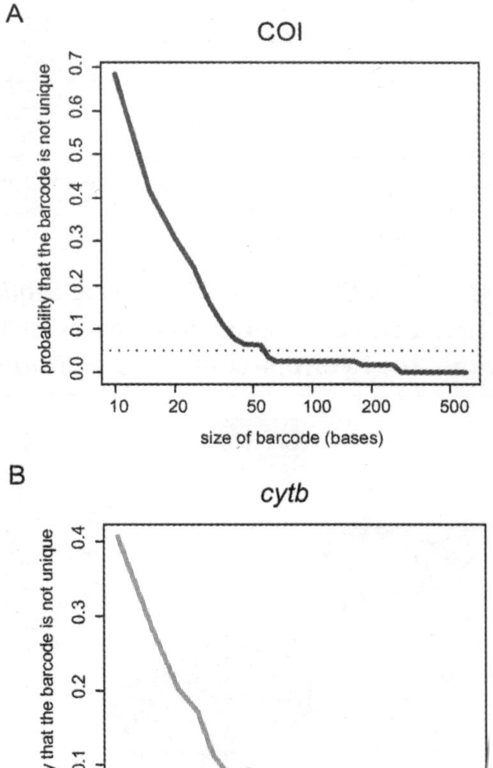

Even 50-base barcodes can discriminate between >95% of species. The analysis is performed by adding sequence information from the 5' region of COI (A) and cytb (B) and calculating the probabilities.

Table 2: COI DNA Barcodes of Varied Lengths are Capable of Identifying Species in Test Assemblage of 840 Individuals from 87 Species of Neotropical Bats

	Length	% Res	% K2P	% Var
COI				
Standard barcode	654	100	20.9	44.5
Mini-barcode 1	109	95.4	23.8	50.5
Mini-barcode 2	109	97.7	17.6	40.4
Mini-barcode 3	109	97.7	22.8	45.9
Mini-barcode 4	109	100	22.7	41.3
Mini-barcode 5	109	100	20.2	45.0
Mini-barcode 6	109	97.7	19.3	46.8

Res, resolution in neighbor-joining analysis (13); K2P, genetic distances based on Kimura two-parameter nucleotide substitution model (20); var, variable sites.

Figure 3: DNA Barcodes of COI and Similarly Sized Sequences from cytb Demonstrate Three Distinct Geographically and Genetically Distinct Groups within the Neotropical Bat Species Sturnira Lilum

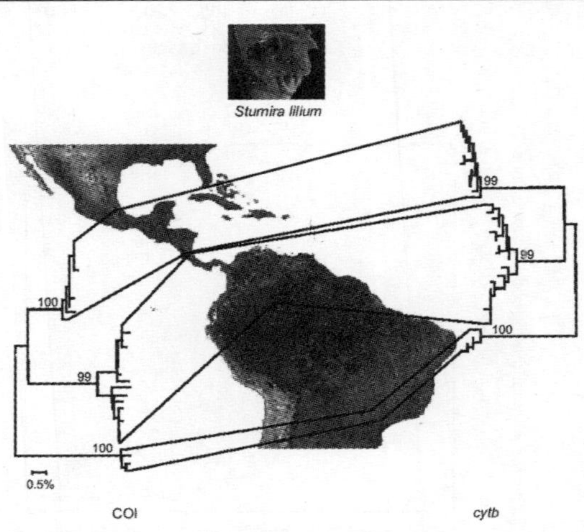

Individual variation between cytochrome b sequences is greater than those within barcode sequences but resolution is internally consistent. The trees are assembled by using K2P genetic distances [20] in a neighbor-joining method [13]. Bootstrap values (1000 replicates) for major geographic lineages are shown above each branch. Photo of Sturnira lilium courtesy of Royal Ontario Museum.

Discussion

This study reveals that both arrays and barcodes are useful tools for the species-level identification of mammals. The main limitation of the array-based approach is that it requires advance knowledge of sequences in target species. Because of a lack of exact matches, undiscovered haplotypes or geographic variants could fail to anneal properly to the probes on the array. While we tried to avoid this problem by providing a set of three different probes per species, this factor can substantially limit the use of microarrays for large-scale biodiversity monitoring. Additionally, to explore unknown species in a given taxonomic group, it might be possible to design probe sets that specifically bind to members of a higher taxonomic level such as genus or family. However, in a situation such as S. lilium, with different geographical variants of up to 8% sequence variation in their COI/cytb genes, a probe set that is designed for the species in one locality might not bind to members of the species in other localities (results not shown). This hit or miss situation could make array technology less desirable in biodiversity monitoring across a wide geographic region. In fact, the current applications of microarrays are usually focused on a limited number of taxa (4). Because of this, assembling a 'Mammalia Chip' might not be a feasible approach for biodiversity monitoring of all mammalian species.

Because barcoding is a sequencing-based technology, it avoids the problem of unknown haplotypes. New haplotypes can be compared to existing databases of barcodes, and they can be assigned to a particular species using probabilistic algorithms (15, 16). The final assignment of a new haplotype to a described species or its assignment to a new species will be achieved through comprehensive taxonomic analysis, which requires different types of data (17). Our analysis supports this argument in all three datasets. While smaller fragments were less powerful in resolving some closely-related species, obtaining more sequence information in these cases (i.e., full-length barcode versus mini-barcode or the whole gene versus the barcode-size fragment) can increase the resolution (10). However, while standard barcode-size fragments (650 bp) can be readily obtained in a single PCR amplification/sequencing from freshly collected or frozen tissue specimens, it is difficult to obtain 650-bp barcodes from specimens whose DNA is degraded (i.e., dried museum samples) (10). The high effectiveness of mini-barcodes means that biomonitoring through barcodes can target different types of specimens, including museum samples or traces of tissues with degraded DNA (10). The mini-barcode strategy also enables exploration of the use of massively parallel sequencing platforms, such as pyrosequencing-based (18) 454 Life Sciences sequencers, for barcoding applications. Interestingly, this technology uses an emulsion PCR approach for simultaneous amplification of several thousand 100-200 base DNA molecules in one reaction. This approach will therefore allow the use of mini-barcodes on environmental samples, which have traditionally been targets for array-based technology.

This study also provides evidence that both COI and cytb are useful species-level molecular markers for mammalian species. This finding is in agreement with earlier work (4). However, when it comes to selecting a molecular marker, it is also important to consider operational issues such as the availability of robust PCR primers, standardization across a wide range of taxa, the robustness of amplifying shorter fragments in PCR reactions of degraded DNA, and the prevalence of mitochondrial nuclear pseudogenes. Our study further confirms that the standard COI barcode can be applied to mammalian species with a similar high species-level resolution as has been observed in other animal taxa tested.

Conclusion

DNA-based methods such as DNA arrays and DNA barcodes provide substantial potential for biodiversity monitoring. However, as the scale of analysis increases, for example in large biodiversity surveys or analysis across wide taxonomic assemblages or different types of specimens, the scalability and sensitivity of these approaches become critical issues in their applicability. Our analyses using three different datasets of mammalian species spanning a wide range of taxa, suggest both DNA arrays and DNA barcodes provide high resolution (i.e., ~95%) across mammalian species. Because DNA arrays might fail to anneal to undiscovered haplotypes of a given species, their use is limited to taxa with known sequences. DNA barcoding, however, provides a higher flexibility for the identification of species in large taxonomic assemblages because it is based on obtaining sequence information that can be used for linking unknown haplotypes to known species.

Methods

Sequence Information

We used COI and cytb genes for array-based and DNA bar-coding analysis of mammalian species by using three taxonomic datasets. The first dataset was selected to allow comparison of the sequence information in the two genes from the same individuals of the same species in a wide taxonomic assemblage of mammals. We used all of the completely sequenced mitochondrial genome sequences of mammals to build this dataset. We downloaded the whole mitochondrial genome sequences of 1585 individuals from 121 mammalian species from GenBank (Additional file 2) and extracted the COI and cytb sequences from them. We refer to this dataset as the mammalian dataset. Our second dataset was selected to test the feasibility of arrays and barcodes in a dense and species-rich neotropical mammalian fauna: 840 individuals from 87 species of bats. This dataset included COI sequences from a recent barcode study on bats (14). We refer to this dataset as the bat dataset. Finally, a third dataset was used as an extension to the bat dataset to compare the utility of both COI and cytb in DNA barcoding of 34 individuals of a single species of bat, Sturnira lilium, from a wide geographic range: 13 localities across nine countries in Central and South America. We refer to this dataset as the S. lilium dataset. Some COI and all cytb sequences for this third dataset were produced in this study (see Additional file 2 for GenBank accession numbers).

Array-Based Analysis

For designing arrays, we chose COI and cytb as separate templates for a probe design algorithm. We assume that probes will be hybridized with amplicons from either COI or cytb of unknown specimens. Our algorithm searched for unique, species-specific sequences, but also considered intraspecific variation among haplotypes of each species (where different halpotypes were available). We designed probes that were 25 nucleotides long and hence suitable for Affymetrix-style single-channel microarrays (Additional file 1). Probes were chosen so that the theoretical probe-target melting temperatures fall within the range of 53.5-58°C, and the GC content falls within the range of 37-54.2%, as recommended by Pfunder *et al.* (4). We designed three probes for each species by using this algorithm. We selected the first 150-bp sequences from the 5' end of each gene as a putative amplicon from which to select the probes.

DNA Barcoding Analysis

For DNA barcodes, we evaluated whole COI and cytb genes as well as various smaller fragments of the two as potential barcodes. For example, we analyzed the whole COI gene of 1557 bases and then performed the same analysis on a 654 base fragment of the 5' region of this gene – corresponding to the standard DNA barcode sequence – as well as smaller, equally-divided 109-bp fragments of the barcode region (i.e., positions 1-109, 110-218 and so on). A similar analysis was performed on cytb by selecting the 5' region of this gene as a potential 654 bp barcode-size region. We used this same analysis for both bat datasets. We counted the number of species with non-overlapping barcodes (i.e. barcodes that uniquely identify individuals of a species) in a Neighbor-Joining (NJ) analysis (13) as a measure of resolution (19). In order to investigate the minimal sequence information required to perform DNA barcoding analysis, we plotted sequence length of putative COI barcodes and cytb gene (sequence information being added incrementally from the 5' end of gene) versus the probability of finding unique barcodes for each species.

Acknowledgements

Funding for this study was provided by Genome Canada (through the Ontario Genomics Institute), and the Gordon and Betty Moore Foundation. Tissue samples of Sturnira lilium were obtained from the Department of Natural History, Royal Ontario Museum, Toronto, Canada. Our thanks to JL Eger, BK Lim and MD Engstrom for facilitating this donation. We thank Daniel Janzen and Donal Hickey for comments on an earlier version of this manuscript.

(Mehrdad Hajibabaei, designed the project, performed DNA barcode analysis, and wrote the manuscript. Gregory A C Singer, gathered sequence information from GenBank, designed and conducted DNA array analysis, and edited the manuscript. Elizabeth L Clare, carried out molecular methods, gathered sequence information of bats, helped with the analysis of barcode sequences, and edited the manuscript. Paul D N Hebert, aided the study design, provided tools/reagents, and edited the manuscript. All authors read and approved the final manuscript. The authors can be reached at Mehrdad Hajibabaei, mhajibab@uoguelph.ca; GregoryAC Singer-gacsinger@gmail.com; Elizabeth LClare-eclare@uoguelph.ca; PaulDN Hebert-phebert@uoguelph.ca respectively.)

References

1. DeSalle R, Amato G: The expansion of conservation genetics. Nat Rev Genet 2004, 5:702-712.
2. Garaizar J, Rementeria A, Porwollik S: DNA microarray technology: A new tool for the epidemiological typing of bacterial pathogens? FEMS Immunol Med Microbiol 2006, 47:178-189.
3. Peplies J, Lachmund C, Glockner FO, Manz W: A DNA microarray platform based on direct detection of rRNA for characterization of freshwater sediment-related prokaryotic communities. Appl Environ Microbiol 2006, 72:4829-4838.
4. Pfunder M, Holzgang O, Frey JE: Development of microarray-based diagnostics of voles and shrews for use in biodiversity monitoring studies, and evaluation of mitochondrial cyto-chrome oxidase I vs. cytochrome b as genetic markers. Mol Ecol 2004, 13:1277-1286.
5. Hebert PDN, Cywinska A, Ball SL, deWaard JR: Biological identifications through DNA barcodes. Proc Royal Soc Lond B Biol Scis 2003, 270:313-321.
6. Hajibabaei M, Janzen DH, Burns JM, Hallwachs W, Hebert PDN: DNA barcodes distinguish species of tropical Lepidoptera. Proc Natl Acad Sci USA 2006, 103:968-971.
7. Hebert PDN, Stoeckle MY, Zemlak TS, Francis CM: Identification of birds through DNA barcodes. PLoS Biol 2004, 2:E312.
8. Ward RD, Zemlak TS, Innes BH, Last PR, Hebert PDN: DNA bar-coding Australia's fish species. Phil Tran Royal Soc Lond B Biol Sci 2005, 360:1847-1857.
9. Hogg ID, Hebert PDN: Biological identifications of springtails (Hexapoda: Collembola) from the Canadian arctic, using mitochondrial barcodes. Can J Zoology 2005, 82:749-754.
10. Hajibabaei M, Smith MA, Janzen DH, Rodriguez JJ, Whitfield JB, Hebert PDN: A minimalist barcode can identify a specimen whose DNA is degraded. Mol Ecol Notes 2006, 6:959-964.
11. Marshall E: Taxonomy. Will DNA bar codes breathe life into classification? Science 2005, 307:1037.
12. Bradley RD, Baker RJ: A test of the genetic species concept: cytochrome-b sequences and mammals. J Mammal 2001, 82:960-973.
13. Saitou N, Nei M: The neighbor-joining method: a new method for reconstructing phylogenetic trees. Mol Biol Evol 1987, 4: 406-425.
14. Clare EL, Lim BK, Engstron MD, Eger JL, Hebert PDN: DNA bar-coding of neotropical bats: species identification and discovery within Guyana. Mol Ecol Notes 2007, 7:184-190.
15. Abdo Z, Golding GB: A step toward barcoding life: A model based, decision theoretic method to assign genes to preexisting species groups. Syst Biol 2007, 56:44-56.

16. Nielsen R, Matz M: Statistical approaches for DNA barcoding. Syst Biol 2006, 55:162-169.
17. Hajibabaei M, Singer GAC, Hebert PDN, Hickey DA: DNA barcoding: how it complements taxonomy, molecular phylogenetics and population genetics. Trends Genet 2007, 23:167-172.
18. Ronaghi M, Uhlen M, Nyren P: A sequencing method based on real-time pyrophosphate. Science 1998, 281:363-365.
19. Hajibabaei M, Singer GA, Hickey DA: Benchmarking DNA bar-codes: An assessment using available primate sequences. Genome 2006, 49:851-854.
20. Kimura M: A simple method for estimating evolutionary rates of base substitution ough comparative studies of nucleotide sequences. J Mol Evol 1980, 16:111-120.

7

Whither 'Community-Based' Conservation?

Chetan Kumar

The concern for biodiversity conservation in recent years has spawned a wide-ranging debate for a return to "fortress" conservation highlighting the inadequacy of community-based conservation. This article examines the nature of the ongoing debate from four key perspectives: incompatibility of conservation and development, viability of strict protection in the current circumstances, moral and economic arguments for conservation and the prevalent orthodoxy of conservation-friendly communities. Despite these limitations, community-based conservation is here to stay. The real issue is not whether communities should be involved, but rather how such involvement can be made effective. Protection of biodiversity must be based on a wide range of approaches to develop a shared understanding of compatible conservation and development goals at various levels.

Source: Economic and Political Weekly, December 2006.

The current concern for conservation of biodiversity in tropical developing countries is fuelled by debates over effectiveness of two mainstream approaches, viz, "fortress" versus community-based conservation. Widespread dissatisfaction of state-led "fines and fences", protectionist, i.e., the fortress approach led to the community-based conservation approach in the late 1980s. It was promoted with the hope of advancing both conservation and development aims, particularly in protected or threatened areas (Wells *et al.,* 1992). Since the 1980s, a myriad of community-based conservation-development initiatives have been implemented throughout tropical Africa, Asia and Latin America. Despite substantial financial investments and efforts, results to date have been mixed – indeed some quite disappointing. Several observers point increasingly to the fragile conceptual and empirical foundations upon which such initiatives are based (Coomes and Barham, 1997; Oates 1999; Barrett *et al.,* 2001).

This article explores both the dominant and commonplace conceptual notions and experiences, which inform the current conservation debate on this issue. It primarily focuses on formal approaches in community-based conservation, which emerged from the 1980s after the western model of conservation based on separation of nature from culture and people, i. e., fortress model did not succeed in protecting biodiversity loss around the world. While attempting to locate the discussions about both approaches and the tensions surrounding them, it argues that implementation of community-based conservation, with the dialectic of a complex, transitional and pluralistic setting is an intricate and in many cases an ongoing process. Therefore, in the current context it may not be the magic portion to cure every conservation problem. But, it definitely is a useful approach for developing a more effective institutional framework to improve conservation-development linkages in several contexts (Hulme and Murphree, 1999).

The article is organised in four sections. The first section traces the history of the two mainstream approaches. Based on this, the ensuing section delves into the arguments of both, the proponents and opponents of community-based conservation over some of the commonly debated issues. It is followed by discussion of some of the major limitations of the community-based conservation approach.

The last section discusses the role community-based conservation should play in the current context. The article concludes that the implementation of community-based conservation programmes is inevitably contentious, and the key issue is to recognise how an effective institutional framework can be developed for a shared understanding of conservation-development goals.

Recognising that the roots of the debate lie in its origin, it is important to understand the history of the two approaches and the whys and the wherefores of the emanating discourse and its consequences leading to the current deliberation.

Foundations of Fortress Conservation

The rise of modern conservation consciousness and conscience, based on the separation between man and nature, dates back to the late 19th century in America and Europe as wilds disappeared and rural communities became urban (Colchester, 1996). In the third world countries, which are the focus of the current debate and this article, conservation action began before the end of the 19th century primarily due to the initiatives of colonial administrators, foresters and scientists (Grove, 1987; Khare, 1998).

The influence of conservation thinking emanating from America and Europe and conservation practices, structures and experiences derived from a range of places from Cape Colony to India (Grove, 1987), towards the end of 19th century, led to creation of areas reserved for game or wildlife. The defining feature of these areas was exclusion and prevention of use by local inhabitants – nature being separate from culture and livelihoods. This setting aside of areas for "nature" or "wildlife", where human use is either prevented or severely constrained is the fundamental philosophy behind the fortress conservation approach (Adams, 2001).

After the second world war, several formal regimes of fortress conservation were initiated in the countries in Africa and Asia. In Africa, the history of creation of game reserves dates back to 1892 (Adams and Hulme, 2001a). The Cape Preservation Act of 1886 was extended to the British South African territories in 1891 (MacKenzie, 1987). Reserved and protected forests were established in India in 1878 and with many years of game laws and their implementation, the National Parks Act was passed in 1934 (Khare, 1998). In Latin American countries, the setting aside of areas for natural reserves was initiated from the

1960s onwards (Barker, 1980). This conservation approach also became a part of the specific international discourse through the work of several influential conservation organisations like the International Union for Conservation of Nature and World Wildlife Fund founded during 1930s (Adams and Hulme, 2001a).

There were variations in interests and objectives in pursing this approach across different countries depending on a complex set of social, political, economic and ecologica lmilieu. But, the common threads were a set of key factors like economic motives for resources appropriation, both for the use of private capital and as a source of revenue for the state (Gadgil and Guha, 1994; Gibson, 1999), a response to increasing concern about environmental degradation and climatic change (Adams 2001), and the romantic vision particularly of Africa as "Eden" teeming with wildlife. This resulted in a broad notion of conservation more as a technical exercise devoid of socio-political considerations (Ellis, 1994).

The legacy of this approach had dominated programmes and polices of tropical biodiversity conservation till the past few decades. The key characteristics of this has been stricter enforcement regulations for exclusion, creation of several new reserves, parks and sanctuaries in different parts of the world at the behest of state governments, foreign and local conservation scientists, research organisations as well as some donor agencies. The justification for persistence and, in many cases, shifting, to centralised management of these areas for conservation was based on threats from deforestation, habitat fragmentation, overkill, secondary extinction and introduced species (Terborgh and van Schaik, 1997), etc. This also included a plethora of other factors ranging from population growth, poverty as well as economic development, political instability, etc (Oates, 1999).

The influence of this development was, however, weakened in the 1980s. This was due to an increasing hostility between local populace and park authorities (Neumann, 1998), declining economic conditions affecting both effective enforcements of conventional conservation and rural poverty (Gibson and Marks, 1995) and overall the increasing loss of biodiversity from around the globe (Wood, *et al.*, 2000). These concerns helped stimulate a massive rethinking of an alternative conservation approach worldwide. The success of some of the community-based conservation projects led to the recognition and preference of community-based conservation over the fortress approach.

Advent of Community Conservation

In the mid-1980s, conservationists began realising that governments lacked resources to conserve biodiversity effectively (Wainright and Wehrmeyer, 1998; Barrow and Fabricus, 2002). This led to the recognition that local communities must be actively involved, and their needs and aspirations considered, if biodiversity is to be conserved (Gadgil, *et al.,* 1993). This wider consensus on community-based involvement was also the culmination of several international initiatives (Adams, 2001). The paradigm shift arose from the recognition that in many parts of the world, conservation was unattainable without the support of the people living in the proximity of the parks and sanctuaries (Ghimire and Pimbert, 1997). These trends led to the development or as some scholars argue the revival of the conservation paradigm of community-based conservation, emphasising natural resource conservation by, for, and with local communities (Western and Wright, 1994; Kothari, 2001).

The fundamental philosophy unlike fortress approach has been that biodiversity conservation will succeed only if local communities receive sufficient benefits, participate in management, and, therefore, have a stake in conserving the resource (McNeely and Pitt, 1987). It also arose in response to criticism levelled at conservationists for the ways they ignored the needs, ideas and aspirations of local people in planning and implementation of programme. Accordingly, implementing organisations were encouraged to deliver community development programmes, promote income generating activities and involve local communities in conservation efforts (Mehta and Kellart, 1998). The community-based conservation paradigm considered conservation and development to be compatible.

With this premise, in the past two and half decades, community-based conservation approaches have proliferated in various parts of the world. A wide range of mechanisms and arrangements for community involvement like Integrated Conservation and Development Projects (ICDPs), protected area outreach, collaborative management, co-management, joint management and community-based natural resource management have been adopted (Barrow and Murphree, 1998).

Framing the Debate

Recent reviews of the community-conservation approach has shown that while the benefits to local people has been mixed, there has been notable lack of successful and convincing cases where there has been tangible improvements in biodiversity conservation (Adams *et al.,* 2004). The varied performance of many of the programmes has led to a resurgence of support for the fortress approach, calling for separation of the development objective, as it undermines conservation efforts (Kramer *et al.,* 1997; Oates 1999).

There are, however, others who believe that community conservation has the potential and that return to fortress conservation would be disastrous (Wilshusen *et al.,* 2002; Berchin *et al.,* 2002).

Based on these reviews, four critical issues are identified, which impinge on these ongoing deliberations more than others. This section synthesises the arguments of proponents and opponents of community conservation on these issues.

Conservation and Development: Are They Compatible?

One of the dominant arguments for community-based conservation emphasises that the standard notion of conservation needs to be reworked. It argues for replacing concepts of wildlife preservation with "sustainable utilisation" thinking in which conservation and development goals are seen as co-dependent. Both the issue of sustainable utilisation and linking of conservation with development mainly in the form of ICDPs have been the focus of debate on compatibility of conservation and development.

In the former case, the origin of the debate dates back to the World Conservation Strategy (WCS) in the 1980s, which promulgated the concept of sustainable utilisation of species and ecosystems (Adams, 2001). Based on the objectives espoused in WCS the proponents argue that conservation should be regarded as an integral part of economic development in poor countries. Whereas, the opponents point out that this presents a simplistic vision of reality and that the mechanisms proposed to attain this goal will irrevocably lead to loss of biological diversity (Robinson, 1993; Oates, 1999). The main critique of this approach has been that sustainable use depletes biodiversity (Redford and Richter, 1999).

However, Wilshusen *et al.*, (2002) point out that the evidence against controlled resource use is not as conclusive as proponents of the protectionist argument suggest. This view ignores social and political realities (i.e., pre-existing use rights) to which interventions must adapt, e.g., it leaves out the fact that, in most cases, parks overlap with or adjoin areas with pre-existing land-use rights.

Another major debate on this issue has been linked to the performance of ICDPs. Over the last decade, there have been over 300 ICDPs world-wide absorbing a major portion of the international conservation funding (Hughes and Flintan 2001). Though it varies within as well as across the countries, the ICDP approach can be broadly defined as one that aims to meet social development priorities and conservation goals (Well *et al.*, 1992). The underlying assumption is that the local people will stop exploiting resources within parks if they achieve increased incomes or are otherwise economically compensated for "opportunity costs". Several reviews of the performance of ICDPs, however, have shown mixed results from around the world. Some have judged it to be "promising but unproven" (Brandon and Wells, 1992; Adams *et al.*, 2004). The IIED (1994) review of some of the well known programmes in Africa points to some gains, but at the same time, there are numerous difficulties in implementing these programmes.

This has led to critiques that question the compatibility of conservation with development (Barret and Arcese, 1995; Noss, 1997). The criticisms have been mainly about "risk-fraught" large investments made in these programmes, which raised unattainable expectations and in many cases development increased human pressures (Wells *et al.,* 1992; Stocking and Perkin, 1992; Southgate and Clarke., 1993).

However, its proponents point out that given the broad range of activities that can come under the umbrella of an ICDP, it is unlikely and undesirable, that ICDP can be categorised as definitely successful or failing (Abbot *et al.,* 2001). Malleson (2002) also points out that such criticism misses the possibility that ICDPs' lack of protection success could be from implementation shortfalls rather than fundamental incompatibility of conservation with development. This also overlooks the impact of intervening variables like conflict, organisation and governance. Hence it would be a mistake to assume out of hand that if ICDPs do not sufficiently address biodiversity protection then we have nothing to learn from the approach and thus it should be tossed out as a policy tool (Wilshusen *et al.,* 2002). Wells *et al.,* (1992) conclude in their review that if the commitment to conserve biodiversity is sincere, then the answer is that ICDP approaches must be reinforced and expanded simply because there are few viable alternatives.

Parks and People: Is Strict Protection a Viable Solution?

Another dominant argument for community-based conservation approach is strongly rooted in the evidence that in all developing countries, local communities continue a day-to-day interaction with the areas and species sought to be conserved; even if not de jure, there is de facto use, even in case of strictly protected areas (Kothari *et al.,* 1998). Across the world half of the protected areas are inhabited (Borrini-Feyerabend 1996). A review of the African situation reconfirms this fact that areas of outstanding conservation coincide with dense human settlement or impact (Balmford *et al.,* 2001). The implicit argument of the proponents, therefore, is that attempts to exclude local people will be unrealistic and will inevitably lead to conflicts and resource degradation. Therefore, participation of local people in management is an imperative (Saberwal., 1996; Brown and Kothari., 2002).

Another important dimension of this debate is the critique by the proponents that protected area management has often been based on a far too static view of ecosystem dynamics. In many cases indigenous or local management practices can provide the controlled disturbance needed to maintain diversity which strict preservation cannot (Ghimire and Pimbert 1997; Gadgil 1998).

This viewpoint has, however, been criticised by Spinage (1998) as misrepresentation of ecological theory and half-truth. The main opposition to this view is that such indigenous management was relevant in the low population and hence low pressure on resources context. Since, over the years human densities have changed around parks, the role of local management practices is overestimated. The opponents argue for no compromise with strict protection (Oates, 1999). As Kramer and van Schaik (1997) recommend, protected areas will always be in need of active defence, no matter how great their benefits are to local communities or to society at large. This argument has been further corroborated by the comprehensive assessments of Burner *et al.* (2001), who found out that parks have been surprisingly effective at protecting ecosystems and species despite significant pressures.

Conservation-Friendly Communities: Myth or Reality?

The proponents of community-based conservation emphasise that the rural or, especially, "traditional/indigenous" communities were in harmony with local resources and have demonstrated long established patterns of sustainable and equitable resource use of these resources (Colchester, 1996). There are numerous studies, which have documented an extensive evidence of patterns of indigenous people having conserved biodiversity (Warren and Pinkston, 1998). Alcorn (1994) argues that this is, however, not recognised by conservationists who wish to avoid evaluation of their own activities. This argument is further pursued by the call for greater recognition of indigenous/traditional knowledge systems in conservation (Gadgil *et al.,* 1993; Kothari *et al.,* 1998). Another strong argument in favour of the presupposition that communities have a greater interest in the sustainable use of resources than does the state or distant corporate managers is that the local communities are more cognisant of the intricacies of local ecological processes and practices. Hence, because of their dependence on these resources in many cases they are more able to effectively manage those resources through local or "traditional" forms of access (Brosius *et al.,* 1998).

However, several scholars point out that the naive assumption of community as a homogeneous group having common vision of conservation is flawed (Agrawal and Gibson, 1999; Leach *et al.,* 1999). The homogeneous images of community often encountered in conservation debate, are a poor reflection of empirical reality and hence often a misleading guide to practical intervention strategies (Li, 1996). There is a need for a critical perspective of "who actually conserves" in various conservation activities. Community-based conservation programmes should recognise the internal differentiation with the communities. Some opponents even question the capacity of communities to self govern the resources at their disposal (Alvard, 1993; Barrett *et al.,* 2001). They claim that in the current context, due to growing population pressure, increased access to modern technology, increasing market orientation, and steady erosion of traditional cultures, there is no guarantee that biodiversity objectives will be achieved if resource control is placed in the hands of local/indigenous groups (Spinage, 1998; Enters and Anderson, 1999). Terborgh (2000) points out that indigenous people, just like people everywhere, act opportunistically in their self interest in exploiting wildlife.

'Fortress' Conservation: Are Economic and Moral Arguments Valid?

The often cited basis for the protectionist approach is based on a combination of economic (pragmatic) and moral arguments (Wilshusen *et al.,* 2002). Kramer and van Schaik (1997) summarise the pragmatic argument by explaining the utilitarian importance of preserving biodiversity in terms of economic use and non-use values. This type of thinking emphasises that humans benefit from wild nature in various ways via ecological services, genetic banks for key agricultural crops, and environmental services etc. Therefore, these benefits should act as a powerful incentive to conserve nature (Balmford *et al.,* 2002). This argument further extends to biodiversity being a common good, and therefore, there are rights of global, regional and local communities to enjoy the aesthetic qualities of nature now and in the future (Terborgh, 1999). In many ways this may seem as valid. However, this line of thinking ignores cultural diversity and how different cultural groups' perceptions of the natural world might affect dialogue on conservation. It also assumes that local and non-local interests are of the same order and carry the same weight and hide the widely held perception that the "common good" refers to elite special interests (Wilshusen *et al.,* 2002).

To summarise the debate based on the arguments presented in the previous sections, greater protection measures cannot safeguard rapidly disappearing tropical biodiversity. Neither can conservation with development be a singular strategy, most likely it will not provide sufficient protection of biodiversity. In such a situation, should community-based conservation be abandoned? This paper argues instead for search for alternatives, it is important to first diagnose the problems of the community-based approach as it still has the potential to address conservation problems.

What Ails Community-Based Conservation?

It is amply clear from the discussion that support for community conservation is by no means universal. The main reasons for this include issues such as the practice of not keeping pace with policy rhetoric, a simplistic understanding of complex intra- and inter-community interactions, inequitable distribution of the rights and responsibilities for natural resource management and losses of power (Barrow *et al.,* 2000). Empirically, there is little doubt that many of the projects implemented with this paradigm shift – including some of those touted as major success stories – are experiencing problems (Berkes, 2003).

One of the key constraints of the community-based conservation approach has been its inability to reconcile the complexity of facilitating development with conservation. The assumed simplicity of linking conservation with development is fraught with problems for both practical and conceptual reasons. These range from lack of delivery of sufficient tangible benefits impacting significantly on people's livelihoods (Inamdar *et al.,* 1999) to the arguments that if conservation is to be based on economic benefits then people who accept it might reject it for a better economic alternative (Hackel, 1998).While project failures could be attributed to several reasons, what is critical is that the links between the two are not predictable (Abbot *et al.,* 2001).

In many cases traditional "community" institutions have been weakened by high levels of political, social and economic uncertainty, and by high levels of population movement. Many communities are now extremely diverse and divided, and often unable to coope2rate internally (Sekharan, 1996). The most daunting problem is that different interest groups subsumed in the category community interact with the local environment and its resources in different ways. These interactions are constantly changing and the challenge of community conservation lies in its ability to cope with this (Enters and Anderson, 1999).

Issues of governance have a strong influence on the conservation, i.e., rules and regulations under which power is exercised for the conservation purposes and the relationships between park managers/government agencies, local people, private sector and even among communities (Browneta1, 2002). The impacts are demonstrated in various forms of conflicts ranging from those emanating due to differing priorities between authorities and local people to managing financial benefits and its distribution (Karlsson, 1999; Songorwa, 1999). Whereas conflicts in the pre-community conservation era took place mainly between communities and conservationists, there is now a new source of tension, within communities. Corruption, nepotism and jealousy raise their heads as soon as community conservation produces meaningful benefits (Barrow and Fabricus, 2002). This is further complicated by the lack of technical and administrative management capacity for an effective monitoring and evaluation of conservation measures and therefore, the scene is set for disillusionment amongst project managers and donors. Added to this is the issue of political struggle over wildlife particularly in many African countries (Gibson, 1999).

Another concern is the fact that many conservation authorities and their technocrats still seem unconvinced of the desirability of building true partnerships with communities. They still view rural communities as technically unable and politically unprepared to play a serious rôle in conservation. In many cases projects have been affected due to the lack of willingness amongst conservation biologists to support the devolution of the control of forest resources to communities and their failure to accept that difficult trade-offs (Malleson, 2002). Little (1994) also cautions about the fact that factors such as the reluctance of the central governments to devolve authority, the difficulty in ensuring compliance with "participation" guidelines, and the large amount of time spent in administration, etc., may work against community conservation in future.

These and many other arguments lead even proponents of community conservation to recognise that community conservation is not a panacea for all environmental and conservation problems (Kothari *et al.*, 1999; Berchin *et al.*, 2002).

Conclusion

Based on the discussions in the previous sections, it is obvious that neither the fortress nor the community-based conservation approach is a uniformly reliable foundation for tropical-biodiversity conservation. The failure of both the approaches has multiple sources. This article has dwelled on four key assumptions informing the debate on current conservation paradigm. It also seeks to highlight that the failure of community-based conservation mainly stems from improper conception and implementation of conservation projects. There is a need for better recognition of the fact that conservation and development cannot be linked in all the contexts.

Hence, the key question which this article seeks to ask is should community-based conservation be allowed to wither? The answer is no! The relevance of the community-based conservation approach lies in the fact that despite its limitations and problems, there is no denying the reality that the ground rules have changed:

no protected area is an island and people and conservation are difficult to separate (Barrow and Fabricus, 2001). As Adams and Hulme (2001b) put it, "love it or hate it, community conservation (in one guise or another) is here to stay". Having said that, it is also imperative to recognise that the real issue is to find ways as to how can it work.

There is a need to build upon the experiences of community conservation approach. One of the key lessons from the community-based conservation approach has been that that biodiversity conservation cannot be done in isolation and that approaches must involve effective partnerships at various levels, i.e., development of new institutional framework (McShane, 2003). Conservation planning without an adequate local partnership is unattainable. However, there is a need to move beyond a single approach to understand what works where and how. As Hackel, (1998) point out, to succeed wildlife conservation policy should be a mix of protectionism, community involvement, public relations, conservation education and revenue sharing.

One possibility as suggested by Barrett *et al.,* (2001) is that the strength of distinct organisations (community, government, NGOs, etc.) should be combined through vertical coordination within nested hierarchies. For example, many communities performing certain tasks in exchange for resources from a few regional governments operating within the bounds of policies established by central governments. Alternatively, horizontal coordination within a level in a hierarchy, such as through federations or union of neighbouring communities around a protected area is another possibility. Berkes (2003) suggest that it may be more useful to rethink community-based conservation as short hand for environmental governance and conservation action that starts first from the ground up but deals with cross-scale relations.

One of the greatest challenges facing conversation today is to be able to engage itself with a broader set of stakeholders and civil society at various levels rather than being confined to debates on success and failures of either top-down or bottom-up approaches. Coalitions composed of conservationists, local people, the forest department and social activists are likely to achieve far more than the current fragmented approach that pits conservationists and foresters against the local community and social activists.

To be effective, programmes for protecting biological diversity must use a wide range of approaches. It certainly starts with more effective participation at all stages of preparation and implementation, from a full range of stakeholders, to reconcile the interest of different interest groups.

There is a need for more flexible designs, improved monitoring and introducing adaptive management to the changing internal and external environments. This calls for more attention to develop an effective institutional framework to engage with a broader set of civil society stakeholders.

The difficulties in putting these suggestions in practice, lie in the fact that it would entail several years of institutional experimentation and adaptation before these initiatives are established. It is evident from the above discussion that conserving remaining biodiversity with the dialectic of complex, transitional and pluralistic settings is an intricate and in many cases an ongoing process. However, given the increasing threats of loss of biodiversity around the world, it is time now to focus on integrating approaches which strengthen the community-based approach.

(Chetan Kumar, Department of Geography University of Cambridge. He can be reached at ck272@cam.ac.uk).

References

Abbot J I O, D H L Thomas, A A Gardner, S ENeba and M W Khen (2001): 'Understanding the Links between Conservation and Development in Bamenda Highlands, Cameroon', World Development, 29 (7), pp 1115-36.

Adams W M (2001): Green Development: Environment and Sustainability in the Third World, Routledge, London, UK.

AdamsW M and D Hulme (2001a): 'Conservation and Community: Changing Narratives, Policies and Practices in African Conservation' in D Hulme and M Murphree (eds), African Wildlife and Livelihoods: The Promise and Performance of 'Community' Conservation, pp 9-23, James Currey, London, UK.

Adams W M and A Hulme (2001b): 'If Community Conservation Is the Answer in Africa, What Is the Question?' Oryx, 35(3), pp 193-200.

Adams W M, R Aveling, D Brockington, B Dickson, J Elliott, J Hutton, D Roe, B Vira and W Wolmer(2004): 'Biodiversity Conservation and the Eradication of Poverty', Science, 306, pp 1146-49.

Agrawal A and C C Gibson (1999): 'Enchantment and Disenchantment: The Role of Community' in Natural Resource Conservation', World Development, 27(4), pp 629-49.

Alcorn J B (1994): 'Noble Savage or Noble State? Northern Myths and Southern Realities in Biodiversity Conservation', Ethnoecologica, 2(3), pp 6-19.

Alvard M S (1993): 'Testing the 'Ecologically Noble Savage' Hypothesis: Inter Specific Prey Choice' by Piro Hunters of Amazonia Peru, Human Ecology, 21, pp 355-87.

Balmford A, J L Moore, T Brooks, N Burgess, L A Hansen, P Williams and C Rahbek (2001): 'ConservationConflictsacrossAfrica', Science, 291, pp 2616-19.

Balmford A, A Burner, P Copper, R Costanza, S Farber, R E Green, M Jenkins, P Jefferiss, V Jessamy, J Madden, K Munro, N Myers, S Naeem, J Paavolla, M Rayment, S Rosendo, Roughgarden, K Trumper and R K Turner (2002): 'Economic Reasons for Conserving Wild Nature', Science, 297, pp 950-53.

Barker, ML(1980): 'NationalParks,Conservation and Agrarian Reform in Peru', Geographical Review, 70 (1), pp 1-18.

Barrett C B and P Arcese (1995): 'Are Integrated Conservation-Development Projects (ICDPs) Sustainable? On the Conservation of Large Mammals in Sub-Saharan Africa', World Development, 23(7), pp 1073-84.

Barrett C B, K Brandon, C Gibson and H Gjertsen (2001): 'Conserving Tropical Biodiversity amid Weak Institutions', BioScience, 51(6), pp 497-502.

Barrow E and M Murphree (1998): 'Community Conservation from Concept to Practice: A Practical Framework', Community Conservation Researchin Africa: Principles and Comparative Practice, Working Paper No 8, Manchester, Institute of Development Policy and Management (IDPM), UK.

Barrow E, H Gichohi and M Infield (2000): 'Rhetoric or Reality? A Review of Community Conservation Policy and Practice in East Africa', Evaluating Eden Series, No 5, International Institute for Environment and Development (IIED), London, UK.

Barrow E and C Fabricus (2002): 'Do Rural People Really Benefit from Protected Areas–Rhetoric and Reality', *The International Journal for Protected Areas,* 12(2), pp 67-79.

Berchin S R, P R Wilshusen, C R Fortwangler and P C West (2002): 'Beyond the Square Wheel: Toward a More Comprehensive Understanding of Biodiversity Conservation as Social and Political Process', Society and Natural Resources, 15, pp 41-64.

Berkes F (2003): 'Rethinking Community-based Conservation', Conservation Biology, 18(4), pp 621-30.

Borrini-Feyerabend G (1996): 'Collaborative Management of Protected Areas: Tailoring the Approach to the Context', International Union for the Conservation of Nature and Natural Resources, Gland, Switzerland.

Brandon K E and M Wells (1992): 'Planning for People and Parks: Design Dilemmas', World Development, 20 (4), pp 557-70.

Brosius J P, A L Tsing and C Zerner (1998): 'Representing Communities: Histories and Politics of 'Community-based Natural Resource Management', Society and Natural Resources, 11 (2), pp 157-68.

Brown D, K Schrekenberg, G Shepherd and A Well (2002): Forestry as an Entry Point for Governance Reform', ODI Forestry Briefing No 1, ODI, London, UK.

Brown J and A Kothari (eds) (2002): 'Local Communities and Protected Areas', Parks: The *International Journal for Protected Areas*, 12(2), pp 1-15.

Burner A G, E G Raymond, R E Rice and G A B daFonseca (2001): 'Effectiveness of Parks in Protecting Tropical Biodiversity', Science, 291, pp 125-28.

Colchester, M (1996): 'Beyond 'Participation': Indigenous Peoples, Biological Diversity Conservation and Protected Area Management', Unasylva,186,47(3),pp33-39.

Coomes, O T and B L Barham (1997): 'Rainforest Extraction and Conservation in Amazonia', *The Geographical Journal*,163(2), pp 180-88.

Ellis S (1994): 'Of Elephants and Men: Politics and Nature Conservation in South Africa', *Journal of Southern African Studies* 20(1), pp53-69.

Enters T and J Anderson (1999): 'Rethinking the Decentralisation and Devolution of Biodiversity Conservation', Unaslyva, 199(50), pp 6-11.

Gadgil, M, F Berkes and C Folke (1993): 'Indigenous Knowledge for Biodiversity Conservation', Ambio, 22, pp 151-56.

Gadgil, M and R Guha (1994): The Fissured Land: Ecological History of India, Oxford University Press, Delhi, India.

Gadgil, M (1998): 'Grassroots Conservation Practices: Revitalising the Traditions' in A Kothari, N Pathak, R V Anuradha and B Taneja (eds), Community and Conservation: Natural Resource Management in South and Central Asia, pp 219-38, Sage Publication, New Delhi, India.

Ghimire K B and M P Pimbert (eds) (1997): Social Change and Conservation: Environment Politics and Impacts of National Parks and Protected Areas, Earthscan, London, UK.

Gibson C C and S A Marks (1995): 'Transforming Rural Hunters into Conservationists: An Assessment of Community-based Wildlife Management Programmes in Africa', World Development, 23(6), pp 941-57.

Gibson C C (1999): Politicians and Poachers, Cambridge University Press, Cambridge, UK.

Grove R H (1987): 'Early Themes in African Conservation: The Cape in the Nineteenth Century' in D Anderson and R Grove (eds), Conservation in Africa: People, Policies and Practice pp 21-40, Cambridge University Press, Cambridge, UK.

Hackel J D (1998): 'Community Conservation and the Future of Africa Wildlife', Conservation Biology, 13(4), pp 726-34.

Hughes R and F Flintan (2001): 'Integrated Conservation and Development Experience: A Review and Bibliography of the ICDP Literature', Biodiversity and Livelihoods, Issues No 3, IIED, London, UK.

Hulme D and M Murphree (1999): 'Communities, Wildlifeand the 'New Conservation' in Africa', *Journal of International Development,* 11, pp277-85.

IIED(1994):'Whose Eden? An Overview of Com-munity Approaches to Wildlife Management', IIED, London, UK.

Inamdar A, H de Jode, K Lindsay and S Cobb (1999): 'Capitalising on Nature: Protected Area Management', Science, 283 pp 1856-57.

Karlsson B G (1999):'Ecodevelopment in Practice – The Buxa Tiger Reserve, the World Bank and Indigenous Forest People in Northeast India Forestry', Trees and People Newsletter, No 38, March, pp 39-45.

Khare A (1998):'Community-based Conservation' in A Kothari, N Pathak, R V Anuradha and B Taneja (eds), Community and Conservation: Natural Resource Management in South and Central Asia, pp 81-101, Sage Publication, New Delhi.

Kothari A, N Pathak, R V Anuradha and B Taneja (eds) (1998): Community and Conservation: Natural Resource Management in South and Central Asia, Sage Publication, New Delhi.

Kothari A (2001): 'Time to Move Out of Africa: A Response to Adams and Hulme', Oryx, 35(3), pp 204-05.

Kramer R A and C P van Schaik (1997): 'Preservation Paradigms and Tropical Rain Forests' in R A Kramer, C P van Schaik and J Johnson (eds), The Last Stand: Protected Areas and the Defence of Tropical Biodiversity, pp 3-14, Oxford University Press, New York, US.

Kramer R A, C P van Schaik and J Johnson (eds) (1997): The Last Stand: Protected Areas and the Defence of Tropical Biodiversity, Oxford University Press, New York, US.

Leach, M, R Mearns and I Scoones (1999): 'Environmental Entitlements: Dynamics and Institutions in 'Community'-Based Natural Resource Management', World Development, 27(2), pp 225-47.

Li TM(1996): 'Images of 'Community': Discourse and Strategy in Property Relations', Development and Change, 27(3), pp 501-28.

Little P D (1994): 'The Link between Local Participation and Improved Conservation: A Review of Issues and Experiences' in D Western, R M Wright and S C Strum (eds), Natural Connections: Perspectives on Community-Based Conservation, pp 347-72, Island Press, Washington DC, US.

MacKenzie JM(1989): 'Chivalry, Social Darwinism and Ritualised Killing: The Hunting Ethos in Central Africa up to 1914' in DAnderson and R Grove (eds),Conservationin Africa: People, Policies and Practice, pp 41-62, Cambridge University Press, Cambridge, UK.

Malleson R (2002): 'Changing Perspectives on Forests, People and Development: Reflections on the Case of the Korup Forest', *IDS Bulletin,* 33(1), pp 94-101.

McNeely J and D Pitt (eds) (1987): Culture and Conservation: The Human Dimension in Conservation Planning, Croom Helm, London, UK.

McShane, T O (2003): 'The Devil in the Detail of Biodiversity Conservation', Conservation Biology, 17(1), pp 1-3.

Mehta J N and S R Kellart (1998): 'Local Attitudes Toward Community-based Conservation Policy and Programmes in Nepal: A Case Study in the Makalu-Barun Conservation Area', Environmental Conservation, 25(4), pp 320-33.

Neumann R P (1998): Imposing Wilderness: Struggles over Livelihood and Nature Preservation in Africa, University of California Press, California, US.

Noss A J (1997): 'Challenges to Nature Conservation with Community Developmentin Central African Forests', Oryx, 31(3), pp 180-88.

Oates J F (1999): Myth and Reality in the Rain Forest: How Conservation Strategies Are FailinginWest Africa,University of California Press, Berkeley, US.

Redford K H and B Richter (1999): 'Conservation of Biodiversity in a World of Use', Conservation Biology, 13, pp 1246-56.

Robinson J C (1993): 'The Limits to Caring: Sustainable Living and the Loss of Biodiversity', Conservation Biology, 7(1), pp20-28.

Saberwal V (1996): 'Pastoral Politics: Gaddi Grazing, Degradation, and Biodiversity Conservation in Himachal Pradesh, India', Conservation Biology, 10 (3), pp 741-49.

Sanderson, S E and K H Redford (1997): 'Biodiversity Politics and the Contest for Ownership of the World's Biota' in R AKramer C P van Schaik and J Johnson (eds), The Last Stand: Protected Areas and the Defence of Tropical Biodiversity, pp 115-32, Oxford University Press, New York, US.

Sekharan N (1996): 'Pursuing the 'D' in Integrated Conservation and Development Projects (ICADPs): Issues and Challenges', Rural Development Forestry Network Paper No 19b, ODI, London, UK.

Songorwa A N (1999): 'Community-based Wildlife Management (CWM) in Tanzania: Are the Communities Interested? World Development, 27(12), pp 2061-79.

Southgate D and H L Clarke (1993): 'Can Conservation Projects Save Biodiversity and South America?' Ambio, 22, pp 163-66.

Spinage C (1998): 'Social Change and Conservation: Misrepresentation in Africa', Oryx, 32(4), pp 265-76.

Stocking M and S Perkin (1992): 'Conservation-with-Development: An Application of the *Conceptin the Usambara Mountains, Tanzania*', Transactions of the Institute of British Geographers, 17, pp 337-49.

Terborgh J and C Pvan Schaik (1997):'Minimising Species Loss: The Imperative of Protection' in R A Kramer C P van Schaik and J Johnson (eds), The Last Stand: Protected Areas and the Defence of Tropical Biodiversity, pp 15-35, Oxford University Press, New York, US.

Terborgh J (1999): Requiem for Nature, Island Press/Shearwater Books, Washington DC, US.
– (2000): 'The Fate of Tropical Forests: A Matter of Stewardship', Conservation Biology, 14, pp 1358-61.

Wainwright C and W Wehrmeyer (1998):'Success in Integrating Conservation Development? A Study from Zambia', World Development, 26(6), pp 933-44.

Warren D M and J Pinkston (1998), 'Indigenous African Resource Management of a Tropical Rainforest Ecosystem: A Case Study on the Yoruba of Ara, Nigeria' in F Berkes and C Folke (eds), Linking Social and Ecological Systems: Management Practices and Social Mechanisms for Building Resilience, University of Cambridge Press, Cambridge, UK.

Wells M K Brandon and L J Hannah (1992): 'People and Parks: Linking Protected Area Management with Local Communities', World Bank, Washington DC.

Western D and R M Wright (1994), 'The Background to Community-Based Conservation' in D Western R M Wright and S C Strum (eds), Natural Connections: Perspectives on Community-Based Conservation, pp 1-14, Island Press, Washington DC, US.

Wilshusen P R, S R Berchin, C R Fortwangler and P C West (2002): 'Reinventing a Square Wheel: Critique of a Resurgent 'Protection Paradigm' in 'International Biodiversity Conservation', Society and NaturalResources, 15, pp 17-40.

Wood A, P Stedman-Edwards and J Mang (2000): The Root Causes of Biodiversity Loss, Earthscan, London, UK.

8

EGenBio: A Data Management System for Evolutionary Genomics and Biodiversity

Laila A Nahum, Matthew T Reynolds, Zhengyuan O Wang, Jeremiah J Faith, Rahul Jonna, Zhi J Jiang, Thomas J Meyer and David D Pollock

Background: Evolutionary genomics requires management and filtering of large numbers of diverse genomic sequences for accurate analysis and inference on evolutionary processes of genomic and functional change. We developed Evolutionary Genomics and Biodiversity (EGenBio; http://egenbio.lsu.edu) to begin to address this.

Description: EGenBio is a system for manipulation and filtering of large numbers of sequences, integrating curated sequence alignments and phylogenetic trees, managing evolutionary analyses, and visualizing their output. EGenBio is organized into three conceptual divisions, Evolution, Genomics, and Biodiversity. The Genomics division includes tools for selecting pre-aligned sequences from different genes and species, and for modifying and filtering these alignments for further analysis. Species searches are handled through queries that can be modified based on a tree-based navigation system and saved. The Biodiversity division contains tools

for analyzing individual sequences or sequence alignments, whereas the Evolution division contains tools involving phylogenetic trees. Alignments are annotated with analytical results and modification history using our PRAED format. A miscellaneous Tools section and Help framework are also available. EGenBio was developed around our comparative genomic research and a prototype database of mtDNA genomes. It utilizes MySQL-relational databases and dynamic page generation, and calls numerous custom programs.

Conclusion: EGenBio was designed to serve as a platform for tools and resources to ease combined analysis in evolution, genomics, and biodiversity.

Background

Large-scale genomic technologies have generated an extraordinary amount of data in the past few decades. Consequently, a huge effort has been made toward creating biological databases and systems to organize, analyze, and share information with the world-wide community (1-4). The application of genomic technologies to molecular evolution has opened new frontiers in the interdisciplinary field of evolutionary genomics, and this has given rise to a great potential to elucidate complex questions in biology (2,5). Understanding of evolutionary processes is critical, since they determine the sequence, structure, and function of macromolecules, and ultimately shape the higher-level biological complexity of organisms.

Genomic biodiversity has been defined as dense sampling of molecular data from diverse taxonomic groups for large genomic regions or complete genomes (6), and inferences concerning evolutionary processes are greatly improved by adopting combined molecular and computational approaches that include a large amount of genomic biodiversity (6-9). We have found that the study of evolutionary genomics in the context of a dense sampling of species gives rise to many unique data processing problems, and so have developed the Evolutionary Genomics and Biodiversity (*EGenBio*) project as a web-based system to simplify large-scale evolutionary data management.

The central aim of *EGenBio* is to provide integrated analysis and visualization of raw sequence data, alignments, and phylogenetic trees, to rapidly curate and annotate that data, and to filter that data based on these annotations for further analysis of specific genomic contexts. It is designed to be robust to change and easily extensible to other data-sets and other analytical programs. To accomplish this, *EGenBio* has web-based interfaces designed to: (1) access computational tools for phylogenetic and evolutionary analyses; (2) facilitate the construction of large-scale sequence and alignment datasets across diverse taxa; and (3) provide a framework for comparative analysis of diverse genes and genomes. *EGenBio* may also serve to promote the utility of increases in the scale of genomic biodiversity.

Overview

The three conceptual divisions of *EGenBio* (Evolution, Genomics, and Biodiversity) serve to organize web-based access to the primary custom-built tools (Figure 1, Table 1). The Genomics division provides access to pre-aligned sequence databases, and serves as a means of flexibly selecting genes and species/genomes of interest and producing a concatenated and annotated alignment for use in further analysis. The data is stored in a MySQL database and accessed "on the fly". Our prototype database contains complete vertebrate mitochondrial genomes that are mostly obtained from NCBI RefSeq annotations (10), but also contains "pre-submission" genomes provided by other investigators. Private access to "pre-submission" genomes allows users to pre-integrate their data with publicly available data for analysis in the primary genome publication, and is available on request. Complex searches can be stored, and registered users can create a personal list of searches that are retained in memory. Alignments for protein-coding genes are based on their amino acid sequences using ClustalW (11), but either amino acid or nucleotide alignments can be selected. The Biodiversity division is a collection of tools for analysis of alignments or sets of sequences (Table 1). These tools do not require a phylogenetic tree. Examples include lists of sequences available, database summaries, graphical representation of gene order information, and summary information on selected or uploaded alignments. This section also includes tools to aid experimental analysis of evolutionary genomics predictions, such as generating primers that represent all possible permutations of a set of sequences, or generating degenerate primers that reflect a sample from a posterior probability prediction of a particular ancestral sequence. The Evolution division includes tools (Table 1) that incorporate or extract information from phylogenetic trees and that translate from accession numbers to human-readable labels for sequences (e.g., genus species designations, or common names of organisms). This division also includes access to programs for processing and visualization of alignment filters (described below), coevolutionary analysis (12,13), and saturation mutagenesis analysis (14).

In addition to the three main sections, *EGenBio* contains a Tools section that serves as a repository for small standalone tools that may also exist as components of other pages, or which serve other simple purposes. The Help link leads not to a separate section, but rather to what we will call a separate "framework". The structure of the Help framework mirrors the main framework exactly, but instead of linking to actual tools, Help pages link to detailed descriptions and documentation for each page. Invisible to most users, a hidden Design framework allows for rapid editing and movement of page and site structure information from design to laboratory testing stages, and finally to public access.

Discussion and Conclusion

EGenBio is a web-based system for analysis in evolutionary genomics and biodiversity. It provides tools and resources for quickly creating, modifying, and analyzing large alignment datasets in ways that we have found useful in our own computer-based and experimental evolutionary genomics research. Our prototype database of complete vertebrate mitochondrial genomes represents the densest complete set of genes currently available from closely related organisms. It can be accessed flexibly according to comma-separated queries or a phylogenetic tree navigation system. *EGenBio* is designed to be easily extensible to use with other protein complexes and other analytical programs. Our goal is to incorporate and utilize as many existing programs as possible, and to develop only "added value" programs. In the current public version, all tools are novel to our system except that alignments are created using ClustalW (11). The PRAED alignment annotation system based on data filters allows alignments to be modified easily according to user interest in annotation features, and allows for the results of analyses to be returned as further annotations on the alignments. Since it is derived from the NEXUS format, it is easy to add batch commands to direct analyses

Figure 1: Organization of EGenBio

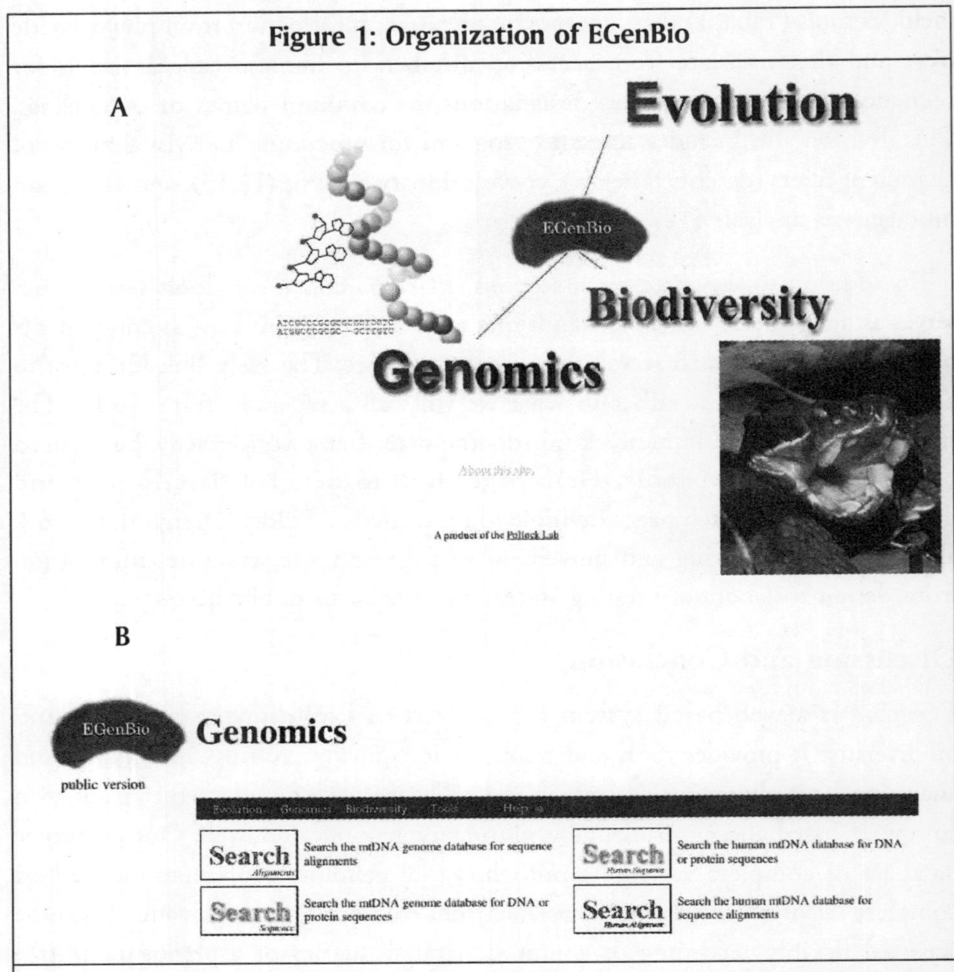

EGenBio is accessed through a splash page that links to the three main divisions, *Evolution*, *Genomics*, and *Biodiversity* (A). Each division has its own access page to the tools that are organized in that division. For example, the *Genomics* division page is shown in (B). The division of each page is clearly marked to allow quick movement among divisions and to the *Tools* section and mirrored *Help* pages.

using many common phylogenetic analysis programs. The PRAED format and data filters are a unique feature of the EGenBio system.

Future modules under development in EGenBio include the creation of additional data filters, incorporation of more genes for analysis of functional divergence, development of further visualization tools for statistical analyses of

Table 1: Custom Tools* Currently in the Main Divisions of EGenBio

Division	Tool Name	Tool Description
Genomics	SearchSequence	Search the mtDNA genome database for DNA or protein sequences
Genomics	SearchAlignment	Search the mtDNA genome database for sequence alignments
Genomics	SearchHuSequence	Search the human mtDNA database for DNA or protein sequences
Genomics	SearchHuAlignment	Search the human mtDNA database for sequence alignments
Evolution	TranslateTree	Translate labels of a tree file
Evolution	FilterViz	Visualize filters associated with alignments
Evolution	TreeReader	Extract tree clusters along with information on branch lengths
Evolution	SaturationTool	Visualize results from saturation mutagenesis MCMC analysis
Evolution	LnLCorr	Detect coevolution among residues using LRTs and trees
Biodiversity	SpeciesList	List species currently in EGenBio
Biodiversity	SpeciesSearch	Search species by taxonomic group or NCBI genome identifier
Biodiversity	LocusOrder	Display mitochondrial gene order for specified taxa
Biodiversity	DatabaseSummary	Provide information about the EGenBio databases
Biodiversity	PrimerPermuter	Generate permutations for use in primer design
Biodiversity	PrimerAlternatives	Produce degenerate primers that reflect amino acid variation

**All tools listed are original to EGenBio, except that alignments are based on ClustalW (11).*

evolutionary dynamics, and automated procedures for analysis using existing programs and tools. We also welcome feedback from the scientific community on areas of general need for integrated evolutionary genomics tools. *EGenBio* is publicly available and can be accessed at *http://egenbio.lsu.edu/ via anonymous login.* User accounts that allow users to save search parameters and results are provided upon request. Incorporation and private access to pre-publication data can also be accommodated upon request. Replication of the *EGenBio* system would require a Linux-based operating system capable of running Perl, Perl-GD, R, PHP, MySQL, and an Apache web server. It would also require installation of numerous custom scripts in addition to ClustalW.

Acknowledgements

This work was partly funded by the National Institutes of Health (R22/R33 Innovation and Development grant to David Pollock), the National Science Foundation (CBM2/EPSCOR), and the State of Louisiana (Biological Computation and Visualization Center, Governor's Biotechnology Initiative, and startup funds to David Pollock). We also anonymously thank other current and former members in the Pollock laboratory for assisting in the development and testing of various tools, and thank Chad Jarreau, Jonathan Bonin, Jonny Roberts Jr., Patricia Ledwig, Stephen McCullough, Sujatha Muralidharan, and Yonatan Platt for contributing to the Biodiversity image collection.

Abbreviations

LRT: Likelihood ratio test; MCMC: Markov chain Monte Carlo; mtDNA: mitochondrial DNA; NCBI: National Center for Biotechnology Information; PRAED: PRagmatic Analysis of Evolutionary Data.

Authors' Contributions

Laila A Nahum is a scientific database curator responsible for the design and documentation of the Help and Design pages and organism database, and co-wrote this manuscript. Matthew T Reynolds, developed and managed the system and has created numerous tools. Zhengyuan O Wang, developed one of the tools and assisted in preparation of the manuscript. Jeremiah J Faith, began initial development of the system and developed several tools. Rahul Jonna, worked on the Help and Design pages, the mirror/framework system, and automated generation of pages. Zhi J Jiang, developed one of the tools and assisted in preparation of the manuscript. Thomas J Meyer, assisted in preparation of the manuscript, and review and editing of the web site, and is developing one of the tools. David D Pollock is the principal investigator responsible for the creation, conceptualization, and management of the EGenBio system, developed early versions of many of the tools, and co-wrote this manuscript. The authors can be reached at Laila ANahum – lnahum@mbl.edu; Matthew T Reynolds – mreyno@lsu.edu; Zhengyuan O Wang – zwang3@lsu.edu; Jeremiah J Faith – faith@bu.edu; Rahul Jonna – rjonna@gmail.com; Zhi J Jiang – zjiang1@lsu.edu; Thomas J Meyer – tmeyer5@lsu.edu; David D Pollock – david.pollock@uchsc.edu respectively.)

References

1. Basu S, Bremer E, Zhou C, Bogenhagen DF: MiGenes: a searchable interspecies database of mitochondrial proteins curated using gene ontology annotation. Bioinformatics 2006, 22(4):485-492.
2. Crandall KA, Buhay JE: Evolution. Genomic databases and the tree of life. Science 2004, 306(5699):1144-1145.
3. Galperin MY: The Molecular Biology Database Collection: 2006 update. Nucleic Acids Res 2006:D3-D5.
4. Vasconcelos AT, Guimaraes AC, Castelletti CH, Caruso CS, Ribeiro C, Yokaichiya F, Armoa GR, Pereira Gda S, da Silva IT, Schrago CG, Fernandes AL, da Silveira AR, Carneiro AG, Carvalho BM, Viana CJ, Gramkow D, Lima FJ, Correa LG, Mudado Mde A, Nehab-Hess P, Souza R, Correa RL, Russo CA: MamMiBase: A mitochondrial genome database for mammalian phylogenetic studies. Bio-informatics 2005, 21(10):2566-2567.
5. Medina M: Genomes, phylogeny, and evolutionary systems biology. Proc Natl Acad Sci USA 2005, 102(Suppl 1):6630-6635.
6. Pollock DD: Genomic biodiversity, phylogenetics and coevolution in proteins. Appl Bioinformatics 2002, 1(2):81-92.
7. Faith JJ, Pollock DD: Likelihood analysis of asymmetrical mutation bias gradients in vertebrate mitochondrial genomes. Genetics 2003, 165(2):735-745.
8. Pollock DD, Bruno WJ: Assessing an unknown evolutionary process: effect of increasing site-specific knowledge through taxon addition. Mol Biol Evol 2000, 17(12):1854-1858.
9. Pollock DD, Eisen JA, Doggett NA, Cummings MP: A case for evolutionary genomics and the comprehensive examination of sequence biodiversity. Mol Biol Evol 2000, 17(12): 1776-1788.
10. Wheeler DL, Barrett T, Benson DA, Bryant SH, Canese K, Chetvernin V, Church DM, DiCuccio M, Edgar R, Federhen S, Geer LY, Helmberg W, Kapustin Y, Kenton DL, Khovayko O, Lipman DJ, Madden TL, Maglott DR, Ostell J, Pruitt KD, Schuler GD, Schriml LM, Sequeira E, Sherry ST, Sirotkin K, Souvorov A, Starchenko G, Suzek TO, Tatusov R, Tatusova TA, Wagner L, Yaschenko E: Database resources of the National Center for Biotechnology Information. Nucleic Acids Res 2006:D173-180.

11. Thompson JD, Higgins DG, Gibson TJ: CLUSTAL W: improving the sensitivity of progressive multiple sequence alignment through sequence weighting, position-specific gap penalties and weight matrix choice. Nucleic Acids Res 1994, 22(22):4673-4680.
12. Pollock DD, Taylor WR, Goldman N: Coevolving protein residues: Maximum likelihood identification and relationship to structure. J Mol Biol 1999, 287(1):187-198.
13. Wang ZO, Pollock DD: Context dependence and coevolution among amino acid residues in proteins. Methods Enzymol 2005, 395:779-790.
14. Pollock DD, Larkin JC: Estimating the degree of saturation in mutant screens. Genetics 2004, 168(1):489-502.

SECTION II

COUNTRY INITIATIVES

9

Gender, Local Knowledge, and Lessons Learnt in Documenting and Conserving Agrobiodiversity

Yianna Lambrou and Regina Laub

This paper explores the linkages between gender, local knowledge systems and agrobiodiversity for food security by using the case study of LinKS, a regional FAO project in Mozambique, Swaziland, Zimbabwe and Tanzania over a period of eight years and now concluded. The project aimed to raise awareness on how rural men and women use and manage agrobiodiversity, and to promote the importance of local knowledge for food security and sustainable agrobiodiversity at local, institutional and policy levels by working with a diverse range of stakeholders to strengthen their ability to recognize and value farmers' knowledge and to use gender-sensitive and participatory approaches in their work. This was done through three key activities: capacity building, research and communication. The results of the LinKS study show clearly that men and women farmers hold very specific local knowledge about the plants and animals

they manage. Local knowledge, gender and agrobiodiversity are closely interrelated. If one of these elements is threatened, the risk of losing agrobiodiversity increases.

1 Introduction

1.1 Biodiversity: Achievements and Challenges Ahead

There is growing worldwide realization that safeguarding the planet's biodiversity is fundamental for agricultural production, food security, and environmental conservation. Genetic resources, because of their diversity, are the cornerstone of sustainable development, as they offer the building-blocks needed to adapt to changing environments and challenges, such as climatic change and increased human pressure on the available natural resources (Gladis, 2003). Many subsistence farmers, especially in environments where high-yielding crop and livestock varieties do not prosper, rely on a wide range of crop and livestock types. This diversity, however, is disappearing at an alarming rate and 75 per cent of today's food is generated from just 12 plants and five animal species. Only 200 out of 10,000 edible plant species are used by humans, and only three plants—rice, maize and wheat—contribute nearly 60 per cent of the calories and proteins obtained by humans from plants. Since the 1900s, farmers have replaced their many well-adapted local crop varieties and land races with genetically uniform, high-yield varieties. Consequently, the small scale and diverse food production systems that conserve crop varieties and animal breeds have been marginalized. Genetic erosion is one of the most alarming threats to world food security. Biodiversity is the arbiter of the quality of human life, and the risk of species loss (Groombridge and Jenkins, 2002), undermines the very sense of 'sustainable development', limits options of the future and robs humanity of a key resource base for survival.

To limit this loss and consequent destruction of natural habitats, farming and land management techniques should be tailored to increase agricultural productivity, while conserving what is left of wild biodiversity. Agricultural policies must change and further action from a range of sectors is needed in the areas of research, public education, development of markets, creation of incentives, implementation of local projects, and investment in ecoagriculture. Special attention should be given to impoverished areas of the biodiversity-rich tropics (McNeely, Jeffrey and Scherr, 2001). To confront the erosion of genetic diversity, Thrupp (2000) proposes the diversification of sustainable agriculture, the use of participatory approaches and building complementarity between agrobiodiversity and habitat conservation in underlying policies.

The three 'Rio Conventions' on biodiversity, climate change and desertification came into existence to highlight the fact that livelihoods and human wellbeing, especially for the poor, are directly threatened by the loss of biodiversity, climate change and increasing desertification. The fundamental interaction between poverty alleviation and biodiversity conservation has already been highlighted in a study by Adams *et al.* (2004) who stress that the Millennium Development Goal (MDG) of environmental sustainability should not be separated from the goal on poverty and reduction of hunger.

The numerous and complex interlinkages between global and local climate, natural habitats and land degradation impact on the rural poor more severely, as they are largely dependent on natural resources for their food security and livelihood. At the global level, deforestation, land degradation and desertification contribute directly to increasing carbon dioxide concentration in the atmosphere; reducing the vegetative cover and impairing the water retention capacity of the soil, and the ability of vegetation to store carbon. Locally, deforestation increases soil erosion, causing a reduction in soil fertility and agricultural productivity. Since forests are the habitat of a large number of species, their degradation results in a direct loss of biodiversity. Land degradation is also a major cause of food insecurity (OECD-DAC, 2001; Lambrou and Laub, 2004).

The loss of both wild and domestic plant as well as animal genetic diversity poses a serious threat to long-term food security. One main threat to the conservation of local farm animal populations appears to be uncontrolled crossbreeding (Wollny, 2003). The maintenance of genetic variation while minimizing counterproductive effects of livestock production on the environment is viewed as a pragmatic and sustainable strategy option, as are the removal of negative economic incentives, improved planning and controlled crossbreeding. A policy promoting decentralized community-based management and full stakeholder participation would alleviate further erosion of the animal genetic diversity.

Biodiversity is critical for minimizing risks in securing rural livelihoods. The reliance of rural women and men on a variety of genetic sources allows them to adapt their agricultural systems to varying environmental, economic and social conditions. It also provides them with a broader income generation possibilities from a wide range of natural resources.

Environmental change challenges the traditional coping and risk-sharing mechanisms based on kin and social groups. If the natural resourcebase is degraded to the point of being insufficient to support the population, drastic measures for ensuring livelihood such as the selling of assets or rural-urban migration are implemented.

1.2 Gender and Sustainable Development

Given the close relationship between desertification, biodiversity erosion and poverty, a gender-sensitive understanding of livelihood roles at the local level is all the more relevant in devising solutions. Women, men, boys and girls perform different tasks that may have direct or indirect effects on the erosion of biodiversity, land quality and water availability. Whatever their roles, the specific targeting of gender and age groups in the assessment of needs, solution design and implementation is an essential factor of programme success. The depletion of natural resources and decreasing agricultural productivity may place an additional burden on women's work and health as they struggle to seek their livelihood in a changing environment. Combined with other pressures, this struggle may subsequently further reduce the time available for women to participate in decision-making processes and income generating activities. Furthermore, climate-related disasters impact more intensely on female-headed households because women generally lack access to, and control over, natural and productive resources (World Bank, 2003a).

Women's participation in biodiversity-related decision-making processes remains limited despite widespread acknowledgement of its importance at the international level. Major obstacles include the lack of secure access to land, adverse financial conditions, public policy traditionally focused on the male population as heads of households, and a strict gender division based along sociocultural norms (Deda and Rubian, 2004).

Local-level biodiversity and environmental integrity are maintained through the long acquired knowledge and experience of both women and men. Such knowledge pertains to domestic plant and animal genetic resources as well as to the quality of soil and water, which form the basis for both the productivity and adaptability of agricultural systems. Wild and semi-domesticated sources offer safety nets in case of food scarcity.

Failure to target both genders in biodiversity conservation and agricultural and rural development initiatives inevitably leads to a loss of knowledge (at local and international levels), and produces a gender bias in policies and programmes (Howard, 2003) which may be detrimental to the functions performed by women. Thus, it is important to empower women and promote an equitable and fair distribution of the benefits and uses of biodiversity (Villalobos *et al.*, 2004).

Clearly, climate change, desertification, and biodiversity erosion have many common causes, and share many elements in terms of adaptation strategies deployed at the individual and policy level. To address the challenges set out at Beijing (1995, 2005) and other international conferences (Cairo, Copenhagen), and to meet the targets embodied in the MDGs, it is crucial to address gender issues in the context of natural resource use and management, particularly as they relate to biodiversity, desertification, and climate change.

Gender equality is vital for achieving all of the MDGs (Grown, Rao Gupta and Kes, 2005). Women's empowerment should be at the centre of development, as they carry the brunt of supporting and caring for families and sustaining life. Practical policies and effective actions should include (i) guaranteed universal access to sexual and reproductive healthcare and rights; (ii) investments in infrastructure to reduce the time and work load of women; (iii) guaranteed property and inheritance rights for women; (iv) elimination of gender gaps in employment and wages; (v) increased political participation for women; and, (vi) combating violence against women. A human rights approach is central to development, with the MDGs and gender mainstreaming as the strategies for achieving human rights (Painter, 2004).

1.3 The International Framework

To analyse the root causes of failed development, UNDP (2003) has examined the structural constraints that impede economic growth and human development, and proposes a policy approach to achieving the MDGs that starts by addressing such constraints. The report proposes more effective aid, new approaches to debt relief, expanded market access to enable diversification and trade expansion, better access to the outputs of global technological progress, follow-through on commitments and setting new targets. Although most solutions to hunger, disease, poverty and lack of education are well known, efforts for their elimination need to be given the proper resources, and services need to be distributed more fairly and efficiently.

Similarly, Oxfam International (2005) called on donors and governments at the 2005 G8 Summit, the UN Millennium Development Goals Special Summit and the World Trade Organization ministerial conference to eradicate global poverty. Oxfam states that the failure to meet the MDGs will cost millions of lives, and failure is in part due to a reduction in the proportion of national spending earmarked to international aid over the past forty years. Oxfam urges world governments to draw up a millennium plan with binding commitments to reform the international trade rules through the cancellation of the debt owed by poor countries, increased volume and effectiveness of aid, to be followed by urgent and concerted action to ensure that commitments are acted upon.

Of particular interest in this regard is the FAO annual report, the State of Food Insecurity in the World (FAO/SOFI). FAO/SOFI (2004) focuses on monitoring the progress towards the World Food Summit (WFS) and MDGs. According to the report, the number of chronically hungry people in the developing world had fallen by only nine million since the WFS baseline period of 1990-92. The WFS Summit goal to halve the number of hungry people by the year 2015 was not only achievable, but also made economic sense. By focussing on simple, low-cost, targeted actions over the next ten years, FAO/SOFI outlines how the resources needed to effectively address food insecurity are very small in comparison to the costs of dealing with the damage caused by hunger. Two parallel strategies are highlighted: (i) intervention to improve food availability and income of the poor by enhancing their productive activities, and (ii) targeted programmes that give direct and immediate access to food to the neediest. SOFI also examines the effect that the rapid growth of cities and incomes in the developing countries and the globalization of the food industry have had on hunger, food security and nutrition.

The interventions and policy measures needed to reduce hunger by half by 2015 have already been identified in a study by Sanchez *et al.* (2005). Concrete steps are proposed in several key areas: (i) investments to improve the agricultural production of food-insecure farmers; (ii) improvements to the nutritional status of the chronically hunger and vulnerable; (iii) investments in productive safety nets; (iv) promotion of rural markets and off-farm employment for increased income; and (v) preservation and conservation of the natural resources essential for food security.

Von Braun, Swaminathan and Rosegrant (2004) suggest that since the majority of poor people rely on agriculture for economic growth, agricultural and rural development is essential to achieve the MDGs economic and social indicators. Strategies should be context-specific, with due consideration to political and economic climate and policy actions, and should create efficient public-private partnerships. Coordination between levels will insure that resources are allotted effectively, and that a sense of ownership is developed with all partners. Nutrition-focused interventions, good governance, and efforts towards peace in conflict-ridden areas must supplement economic growth. Policy action in the critical areas of sustainable agriculture and food nutrition and security is essential for responding effectively and responsibly towards reaching the MDGs (Von Braun, Swaminathan and Rosegrant, 2004).

Kameri-Mbote (2004) examines the implications of international agreements on land and resource rights as they relate to access, control and ownership. He points to agreements that have promoted as well as hindered enjoyment of land and resource rights at different levels, with particular attention to the Convention on Biological Diversity (CBD). The CBD is riddled with contradictions as it tries to accommodate access to resources that must be shared equitably between developed and developing countries. Kameri-Mbote concludes with an examination of the wider context, and highlights the Pan-African Programme on Land and Resource Rights as a way to optimize the benefits of international agreements in realizing land and resource rights for the poor.

With regards to the issue of environment and sustainable livelihood, literature discusses the implications and constraints for sustainable development. Spangenberg (2002) examines sustainability indicators, recommending the inclusion of gender issues, labour, the environment and economy, and peace. Dovie (2002) investigates the link between Agenda 21 and sustainable livelihoods, pointing out that institutions implementing Agenda 21-related activities have often concentrated on economic development at the expense of the environment and poverty reduction in the south. Barber (2003) analyses the elimination of unsustainable production and consumption as one of the three objectives of sustainable development debated at the World Summit on Sustainable Development (WSSD). A global strategy to achieve sustainable production and consumption would come not from a UN consensus of world leaders, but rather through the strategic alliance of responsible governments, civil society and others with a vision beyond the next election cycle.

The International Treaty on Plant Genetic Resources (ITPGR) for Food and Agriculture has generated an interesting discussion. Fowler (2004) looks at the Multilateral System of the International Treaty on Plant Genetic Resources for Food and Agriculture (PGRFA), and analyses the key ambiguities and problems in the text. Details cover the status, scope, major provisions of the PGRFA, the multilateral system and crops, and the relevance of Consultative Group on International Agricultural Research (CGIAR) collections. Ambiguities, on the other hand, include the lack of definitions of important terminology; the Treaty's exemption for facilitated access to material 'under development'; the lack of specificity for benefit-sharing provisions; and stipulation of responsibilities by governments towards PGRFA. Although the Treaty provides a formal framework that clarifies many issues on sustainable agriculture, the issue of farmers' rights is side-stepped, leaving further clarification to individual nations. Despite its potential shortcomings, the Treaty provides a medium in which trust can grow, and implementation must be considered. Cooper (2002) also analyses the main features of the ITPGR, reviewing some of the key negotiation issues and its relationship to the CBD.

1.4 Local Partnership

Roe (2004) looks at poverty, environment and the achievement of the MDGs through an integrated approach to conservation and development. He argues that ecosystems have to remain intact as a basic human requirement, and that communities and local partnerships are a vital force to sustainable development. He emphasizes increased awareness amongst development agencies about the importance of conservation, by recognizing and strengthening the comparative advantage that biodiversity offers to many poor countries. A shift in the focus of international conservation policy—from looking primarily at rare and endangered species towards emphasizing the development values of biodiversity and landscape management approaches—is necessary.

The management of local resources has a greater chance of achieving a sustainable outcome when a partnership exists between the local people and external agencies. (Pound *et al.*, 2003). As indicated in the participatory approach, in order to improve natural resources management, it is necessary to incorporate participatory and user-focused approaches that lead to a development model based on the needs and knowledge of local resource users. Such an approach is also recommended by Ramírez and Quarry (2004), who draw particular attention to the importance of exchanging knowledge and information, and developing awareness.

Laird (2000) offers practical guidance on conducting equitable biodiversity research and prospecting partnerships. These recommendations include developing research codes of ethics, designing effective commercial partnerships and biodiversity prospecting contracts, and drafting and implementing national 'access and benefit-sharing' laws, combined with institutional tools for the distribution of financial benefits.

Calderón (2004) points to the need to move beyond the top-down charity approach and project development models to models that are based on collaborative action for social change. The value of participatory research—utilizing traditional farmer knowledge—has already been highlighted by Goma *et al.* (2001) in a discussion on the relevance of an interactive farmer-researcher process.

Participatory research is expected to improve the efficiency, equity, and sustainability of natural resource management research and development (R&D) projects by ensuring that research reflects users' priorities, needs, capabilities and constraints. Particular attention should be given to contributions from women and other marginalized groups (Johnson *et al.*, 22004).

Community-based, participatory and co-management processes are often slower and more complex than traditional bureaucratic or technical project implementation, however, participation and at least partial control over the process from research to implementation and beyond is seen as central to an effective empowerment strategy (Simon *et al.*, 2003).

Vernooy (2003) encourages collaboration between researchers and farmers, as participatory plant breeding is instrumental for the development of plant varieties that truly meet farmers' needs. He examines research questions, the design of on-farm research on the rights of farmers and plant breeders, and argues for the development of new supportive policies and legislation. Vernooy recommends action to ensure that participatory plant breeding achieve the intended results. Maier (2002) calls for an international convention and treaty on livestock genetic resources to establish legal recognition of the rights of pastoralists and livestock keepers.

Investment in research is thus crucial. However, in the last decade or so, there has been a decline of public investments for research, especially in Africa, and funding has become more donor dependent. Although the efficacy of donor-supported projects has helped to build capacity in many countries, advances can be quickly eroded if donor funding is withdrawn and other sources are not consolidated or developed further (Beintema, Nienke and Stads, 2004).

1.5 Farmers' Rights

Borowiak (2004) examines the rights of farmers as a resistance strategy against the perceived inequities of intellectual property rights regimes for plant varieties. The campaign to legitimize the traditional seed-saving practices of the farmers alongside the increasingly commercial models of intellectual property in agriculture had mixed implications. Borowiak admits that this campaign could help transform conventions of intellectual property to become better suited for registering and for providing financial encouragement to alternative forms of innovation. However, the enactment of farmers' rights has been difficult. By comparing the rights of farmers to those of commercial breeders, Borowiak cautions that the campaign risks further legitimization of inequities, favouring the interests of the seed industry to the detriment of the farmers.

Srinivasan (2003) examines the feasibility of the provisions on farmers' rights in plant variety protection legislation. He argues that the provisions by some developing countries will involve substantial operational challenges. IPR-based farmers' rights are unlikely to provide significant economic benefits to farmers and their communities, as these are not likely to diminish the incentives provided to institutional plant breeders. Indian Plant Variety Protection (PVP) legislation is used as an example, as this appeared to have gone quite far in articulating the provisions on farmers' rights. Conservation projects supported by community gene funds are a more efficient way to preserve agrobiodiversity than extending the IPR regime to farmers' traditional varieties. He cautions that the resources recuperated from breeders' IPR research may not be adequate to realistically fund this strategy.

Similarly, Brush (2005) questions the significance of bioprospecting in protecting traditional agricultural knowledge and argues for a common pool approach with genetic resources remaining in the public domain. Brush examines the nature of crop genetic resources, farmers' knowledge, and the nature of the 'common heritage' regime that was being partly dismantled by the CBD. He reviews the implementation of access and benefit-sharing schemes under the CBD and discusses programmes to recognize farmers' rights that have arisen since the establishment of the Convention. He argues for increased development assistance to be focussed on programmes for improving rural income in genetically diverse farming systems. The challenge to establish farmers' rights should follow from the International Treaty on Plant Genetic Resources for Food and Agriculture (ITPGRFA) and India's Act 53, which emphasize multi-community solutions rather than individual contracts for accessing crop resources and sharing benefits from their use.

1.6 Local Knowledge

Although traditional farming systems are diminishing worldwide, their role remains crucial for maintaining community food security and for conserving agrobiodiversity, as well as for the design of more sustainable agroecosystems appropriate for small farmers (Altieri, 2004).

With particular regard to natural resources management, the conventional 'indigenous knowledge' approach shows that shortcomings can be circumvented through a subset approach called 'traditional ecological knowledge', which adds an explicit ecological emphasis to the conventional development method (Dudgeon and Berkes, 2003).

UNESCO (2002) points out that insufficient attention has been paid to the relationship between indigenous knowledge and power, and they advocate increased attention to be focused on the context within which indigenous peoples live. Particular attention should be paid to political relations. It is important to develop a relationship between the scientific community and the holders of traditional knowledge. This calls for a more equitable partnership that fully respects indigenous peoples, their territories, and self determination (ISCU, 2002).

Until the CBD recognizes the existence of indigenous peoples and the rights of indigenous peoples as set out under international law, the promise of the Convention is likely to remain unfulfilled (Oldham, 2002). Ruiz (2004) discusses traditional knowledge as a tool, which enables the countries of origin to assert their rights over their genetic resources, to benefit from such resources in an equitable manner and to protect indigenous peoples' intellectual efforts.

Some authors discuss local knowledge from the seed systems' view. Tripp (2000) analyses the inability of formal African seed systems to meet farmers' needs and suggests that such systems could be strengthened in countries such as Kenya, Malawi, Zambia and Zimbabwe by considering the nature of seed demand, provision, and emergency distribution programmes, as well as policy and regulatory frameworks, and the role of public sector research. He recommends that precise national strategies be developed, and a sustainable seed system be created as a combined effort of public, commercial and local-level stakeholders. He recommends seed policy reform in much of Sub-Saharan Africa. Similarly, Louwaars (2000) points to the risk of introducing seeds regulations, which are often inappropriate for the local informal seed systems that have taken generations to evolve. Such regulations could restrict informal seed systems and in some cases the initiatives by local farmers could be construed as illegal, limiting recognition and reward from these systems.

1.7 HIV/AIDS, Food Security and Biodiversity

Gillespie *et al.* (2001) examines how HIV/AIDS affects nutrition, food security, and household livelihoods, and those dependent on agriculture. They discuss mitigation as the primary public sector response to these challenges, suggesting that key generic public policy and programming principles should include 'doing no harm'. These authors suggest mainstreaming HIV/AIDS concerns into food and nutrition programming, with due consideration to scale, context, targeting, monitoring and collaboration. They conclude with a call to re-examine policy.

Jayne *et al.* (2004) propose modifications to existing agricultural policies and programmes for better achievement of policy objectives in the context of the HIV/AIDS epidemic in eastern and southern Africa. The effects of the epidemic are likely to affect the agricultural sector in numerous ways, increasing the cost of labour and scarcity of capital. They suggest improvements to technical capacity; rehabilitation of agricultural extension services and institutions for the crop and input marketing systems that contribute to small-scale farmers' productivity and food security.

Kengni *et al.* (2004) examine the potential of forests to provide food security for resource-poor rural families against the socioeconomic impact and livelihood threats from HIV/AIDS. They analyse the role of local food-based approaches in rural communities where short- and long-term goals are maintained and food security needs are met while preserving the natural resource base and conserving indigenous fruit and vegetable species. According to the study, wild foods can be cheap, nutritious, and economically beneficial, and their production can be less labour-intensive. Kengni *et al.*, conclude that wild foods may provide an alternative to the food shortages and income problems caused by HIV/AIDS if existing added-value technologies are improved and made available to the farmers at low cost.

Barany *et al.* (2001) highlight the contribution of forests to household nutrition and health. They draw attention to the gap in literature on the importance of forest-based research in connection with coping strategies for mitigating the socioeconomic impact of HIV/AIDS on rural agrarian households. The strong traditional dependence of local people on forest resources for health and nutrition could be made compatible to agroforestry systems by taking into consideration the productive challenges associated with low household labour supplies.

Gari (2002) explores the strategic components of the agricultural sector's response to food insecurity and the impact of HIV/AIDS on rural development in Sub-Saharan Africa. He discusses agrobiodiversity and its close relationship to indigenous knowledge as well as their often overlooked albeit important roles in enhancing food security in rural communities affected by the epidemic. He states that the promotion of agrobiodiversity and indigenous knowledge represents a renewed emphasis on local resources and the ability to strengthen agriculture, food and health. Gari recommends immediate and urgent participatory and grassroots-oriented research and action.

McMichael (2004), on the other hand, traces the history of the emergence of new or unfamiliar infectious diseases, HIV/AIDS included. The rise of modern medicine and other rapid changes in demography, environment, behaviour and technology in the human ecological system have also contributed to a rise in biodiversity. He urges for greater understanding of the dynamic process of viruses and diversity in order to anticipate an amoral, self-interested co-evolutionary struggle. Such an understanding can influence environmental management, poverty alleviation, help to reduce susceptibility to disease, foster social capital, and limit ecological damage arising from consumer or commercial incentives, as well as restore society's public health capacity and function.

1.8 FAO and the Global Challenge

The particular role of FAO in the establishment and management of the plant genetic resources for food and agriculture is examined by Andersen (2003). He reviews the main achievements and limitations, with particular focus on the FAO's Commission on Genetic Resources for Food and Agriculture (CGRFA). He examines the CGRFA and its role in the implementation of the International Treaty on Plant Genetic Resources for Food and Agriculture (2001). FAO plays an important agenda-setting function with regards to the sharing of genetic resources and information at the international level, providing an arena for discussion. With strong political support for implementation and funding, the

Box 1: Incorporating Gender-Sensitive Approaches into Plant Genetic Resources Conservation – A Global Workshop

In October 1996, in partnership with the International Plant Genetic Research Institute (IPGRI), FAO brought together various experts focusing on gender and agriculture as well as experts from plant genetic resources to address gender concerns in relevant international policy frameworks and agreements on Plant Genetic Resources (PGR) and the implications for rural women and PGR.

PGRFA can be one of the key elements in halting genetic erosion and providing access to the remaining genetic resources essential for future food security.

Because of its mandate, FAO has long been involved in assessing and addressing environmental and natural resources issues, the interactions between people and the resources around them, and the interlinkages between gender, poverty, agriculture and food security. FAO's current gender and development plan of action specifically recognizes natural resources as a priority area for gender mainstreaming. FAO has long recognized the strong linkages between the gendered knowledge and skills and biodiversity so critical to agricultural production and food security, and has supported various initiatives to this end in Asia, Africa, and Latin America. Given its mandate within the UN system for food and agriculture, it has an important role to play also at the level of international policy. To this end, much support was provided through the 1990s to make sure that gender remained on the international agenda, particularly in arenas such as the Commission on Genetic Resources for Food and Agriculture (CGRFA); particular attention was given to the gendered dimension of farmers' rights.

Through its many initiatives, FAO has reaffirmed that in order to promote and ensure sustainable use of resources and sustainable agricultural development, it is crucial to begin any agricultural planning or policy development process with a gender-sensitive participatory identification of the issues. Following on this, FAO has provided support to member nations in the design of agricultural policy in the effort to promote more gender equitable development. Furthermore, FAO has also extensively supported training efforts in member nations to increase the capacity of national partners to undertake agricultural initiatives in ways that support more sustainable practices through gender-sensitive participatory assessments and development.

Recently, FAO undertook a study to assess the gendered dimensions of the three Rio Conventions on biodiversity, climate change, and desertification (Lambrou and Laub, 2004). In 2005, it also produced a paper that considers gender as the missing component of the response to climate change (see Piana and Lambrou, 2005). These efforts contribute to the international debate of the three environmental conventions. FAO has raised the profile of the link between gender equality and several natural resource concerns related to dryland management, freshwater use, land tenure and property rights, and the need for gender-sensitive indicators at several levels—from the national to the project level—to monitor progress towards sustainable use of natural resources.

2. FAO Project Case Study: Gender, Biodiversity and Local Knowledge for Food Security

The FAO regional project 'Gender, Biodiversity and Local Knowledge for Food Security' evolved because of the growing interest and recognition that rural men and women have in-depth knowledge and understanding of local ecosystems and environmental processes. The aim of the LinKS project was to improve rural people's food security and promote sustainable management of agrobiodiversity by strengthening the capacity of institutions to utilize in their programme and policies participatory approaches that recognize the knowledge of male and female farmers. This section of the paper highlights some of the project's experiences with the participatory approaches to improve people's livelihoods in the long-term.

First, the international context of the project is presented to show how it tried to respond to debate and issues at the global level. Second, there is a brief presentation of the project and its activities. Project accomplishments and challenges are then described, as are the project responses to the challenges it faced. Finally the way forward after the conclusion of the project is described.

2.1 Why this Project?

The International Context

The LinKS project was conceived during 1994-96. In the early 1990s, important international debates focused on the sustainable management of natural resources, biodiversity and participatory approaches. The LinKS's conceptual framework clearly reflects these issues.

In the period leading up to 1996, the understanding of gender, local knowledge systems and the rich source of information embodied in the knowledge, skills and practices of women as managers and users of diversity was not very clear. During the International Technical Conference on Plant Genetic Resources for Food and Agriculture held in Leipzig in 1996, these issues were given greater importance than during the formulation process of the CBD and the Agenda 21

(1992). The CBD addressed the issue of local knowledge in two Articles, 8(j) and 10(c). However, both articles were relatively vague. The Leipzig conference approved the Global Plan of Action, which sets the stage for the development of mechanisms and programmes to be carried out at policy, institutional and community levels to ensure the conservation and sustainable use of plant genetic resources. It also highlights the importance of men and women farmers and their role and contribution to the sustainable management of plant genetic resources.

A joint workshop organized by IPGRI (International Plant Genetic Resources Institute) and FAO on how to incorporate gender-sensitive approaches in the conservation and utilization of plant genetic resources was organized after the Leipzig conference. This workshop was one of the first attempts after Leipzig to link policy with practical activity in the field, activity in which women farmers and resource managers play a crucial role.

After extensive negotiations, the International Treaty on Plant Genetic Resources (ITPGR) became effective in June 2004. The ITPGR, responding to outstanding issues not covered by the CBD, was an important breakthrough, as it formally endorses farmers' rights through a legally binding instrument at the global level. Farmers' rights, based on the recognition that farmers play a crucial role in the management and conservation of plant genetic resources, include the protection of traditional knowledge, participatory decision-making and the right to equitable participation in sharing the benefits arising from the utilization of plant genetic resources for food and agriculture.

In recent years there has been a proliferation of international fora considering different aspects of the protection of the technology and knowledge of indigenous people and local communities. The following discussion demonstrates how the project responded to the discussions, international conventions and treaties.

The LinKS Project

The LinKS project evolved because of the growing conviction that rural men and women have detailed knowledge and understanding of local ecosystems and environmental processes. Furthermore, rural people's traditional practices and knowledge systems are at risk of being marginalized and lost. Thus, the project's goal was to increase among development practitioners the understanding of the value of this knowledge base—and how it can be applied to support valid systems of managing the environment, farming and producing food—for the ultimate benefit of men and women farmers. To achieve its goal, the project initiated and supported partner organizations' activities in three major areas: (i) capacity-building; (ii) research and documentation, and (iii) communication and policy debate.

There is a great overlap between these components, and the activities represent a set of interactive, inter-related and mutually reinforcing processes of support to partner organizations. The project sought to explore these issues with a diverse group of organizations and individuals in four countries in the Southern African Development Community: Tanzania, Zimbabwe, Mozambique and Swaziland. The main strategy of the project was to support, build on, and strengthen the efforts of other groups already working on food security, indigenous knowledge and agrobiodiversity issues in the four countries. These included NGOs, research, training and academic institutions, and government agencies and policy institutions. The project focus was participatory, which meant that project teams and management promoted participatory principles and approaches in actual project management as well as in its activities.

The following sections describe the accomplishments of the project and summarize the important lessons learned from project implementation.

3. Accomplishments of the Project

The first activities within the project began in 1997 in Zimbabwe and Tanzania in two phases: 1997-2002 and from 2002 to September 2005. A stakeholder analysis was carried out in each country, which revealed that partners and important actors in the agricultural sector lacked a common understanding of the issues of gender, biodiversity and local knowledge, and how these relate to food security. Partners hoped that the project could provide opportunities for learning about the issues. Thus as a preliminary step, strong focus was put on raising awareness and strengthening capacity among development practitioners in order to meet the project's ultimate goal of promoting approaches that recognize the importance of men and women farmers' knowledge for the sustainable management of agrobiodiversity and enhanced food security.

3.1 Capacity-Building

LinKS placed a strong focus on institutionalization and the uptake of gender-sensitive and participatory approaches to biodiversity conservation in institutions' on-going programmes. Priority was given to institutions that were able to apply these approaches in their programmes targeted towards rural men and women farmers. This included agricultural extension services, development projects, NGOs and institutions of higher learning.

Several workshops were organized to document traditional practices and to address the main challenges and constraints. Two main issues were emphasized: first, the potential benefits and risks of sharing such knowledge, and second, the responsibilities of researchers and development agents to record and document local knowledge. An attempt was made to develop a set of simple basic guidelines for all involved in the documenting and sharing of local knowledge (FAO, 2000: 7).

FAO carried out several missions to Swaziland to meet various potential project partner institutions, as well as government agencies and civil society organizations to assess the needs and interests of a diverse group of Swazi partners for intensifying activities in the country. After initial consultation and exploratory period, it was concluded that there were already various local initiatives addressing the issue of local knowledge for food security, biodiversity management and integration of gender concerns in agriculture sector. Thus, it was recognized that the time was ripe to broaden project activities to support such local initiatives. Several specific training workshops were organized, and these attracted considerable interest from different organizations.

The main objective was to strengthen knowledge and skills in implementing gender-sensitive participatory agricultural/livestock research and training so that local knowledge systems in agrobiodiversity management for food security would be understood by all participants.[1]

Other workshops were also held that sought to strengthen knowledge and skills in implementing gender-sensitive research to agriculture/livestock so as to better understand the vital role played by local knowledge systems in agrobiodiversity management for food security. The team in Tanzania learned

---

1 See documentation of the 9th LinKS training workshop (FAO, 2005).

how to improve and strengthen the capacity-building component of the project; how to enhance the capability of participants in collecting, documenting and sharing local knowledge related to agrobiodiversity conservation and food security within the framework of their institutions; and were provided insights on the preparation of draft guidelines for the documentation of local knowledge (FAO, 2005). Some other relevant capacity-building activities covered:

- 1125 people participating in the training workshops on gender, local knowledge and biodiversity and the application of gender analysis and participatory methods. More than 830 people enhanced their understanding of the relationship between gender, local knowledge and biodiversity as well as of national and international policy frameworks by taking part in seventeen project-supported thematic workshops and seminars. Most of the participants were involved in research on LinKS issues, enabling them to further enhance their skills and understanding.
- A training manual, building on local knowledge, gender and biodiversity, was developed, highlighting the specific concepts and links between these issues from the perspective of a sustainable livelihoods: see *www.fao.org/sd/LINKS/documents_download/Manual.pdf*
- A local pool of experienced and well trained trainers was built up, to facilitate with the training workshops on LinKS issues and gender-sensitive participatory approaches.
- Redesign of the existing curriculum was undertaken, through project support, to mainstream and institutionalize LinKS issues in training colleges, universities and other institutions of higher learning. FAO supported the workshops on mainstreaming LinKS issues at the Sokoine University of Agriculture. Visits were organized to provide farmers, researchers, NGO representatives and development workers with an opportunity to exchange ideas and experiences, and to take part in mutual learning experiences. In Tanzania, for instance, as part of a research project focusing on the management of animal genetic resources by the Maasai, pastoralists from various study areas exchanged visits to share experiences and views.

3.2 Research

The main rationale for the support of research activities was to develop a better understanding of the linkages between LinKS issues, and to reinforce collaboration between researchers and rural communities; to demonstrate the complementarities between the local and scientific systems of knowledge, and to enhance the potential of developing approaches to increase food security and agrobiodiversity. Research activities were closely linked to capacity-building and advocacy, as they were seen to be mutually reinforcing. Government officers, researchers and NGO staff who participated in the training and awareness workshops, often developed research proposals for increasing recognition of the knowledge of men and women, the documenting of experiences, for community-to-community exchanges, or for follow-up action. All research activities explored the hypothesis that women are important custodians of knowledge in the management of biodiversity.

The stakeholders identified three broad topics as particularly important: (i) traditional seed systems; (ii) animal production and genetic diversity, and (iii) the relation between HIV/AIDS and local knowledge systems.

In total, twenty-eight research activities focusing on gender, local knowledge and agrobiodiversity were implemented.

Traditional Seed Systems

Research activity on gender biodiversity was set up in the southern highlands of Tanzania, a region that has been heavily exposed to seed interventions, thus increasing the availability of improved varieties. The overall goal was to improve the availability and accessibility of high-quality seed of crop varieties preferred by farmers, thus enhancing household food security. At the end of the project, the following main findings were noted: (i) some crop species had disappeared due to changes in weather, migration, government policies and interventions, or farmers' preferences, but at the same time, many varieties with different characters had been introduced, increasing agrobiodiversity; (ii) in general, agrobiodiversity had increased over the years; (iii) levels of food consumption and their composition varied within the different socioeconomic groups; (iv) food-secure households relied more on staple food and less on natural and collected crops; (v) the informal system was a better source of seeds and information for many farmers than the formal seed system. In the Malinzanga and Shinji villages, HIV/AIDS had affected food and seed security in the afflicted households, because of diminishing labour, increasing number of dependants/orphans and weakening physical state due to the illness. The number of female-headed households in the villages affected by HIV/AIDS also increased (Mkuchu, 2006).

Animal Production and Genetic Diversity

In the Mbarali district, a study was conducted to gauge local knowledge on breeding and selection of livestock in the Maasai community, by examining the types of animals (cattle, sheep, goats) preferred and what were the criteria used to achieve the desired traits. These preferences were analysed in relation to gender and age, roles and responsibilities, decision-making, goals of food security and herd survival. The objective was to let the Maasai pastoralists identify the gaps and make corrections. The threats or constraints to the pastoralists' local knowledge for the sustainable management of indigenous livestock were identified, and possible solutions offered. The decreasing grazing land and water for livestock in Mbarali district, and livestock diseases were major constraints.

The Relation between HIV/AIDS and Local Knowledge Systems

A study on the impact of HIV/AIDS on local seed systems in both Tanzania and Mozambique showed that local knowledge is gender specific. Men and women are responsible for different crops; for example, a widower would not necessarily know or be able to produce, after his wife's demise, the local crops she had planted. Her specific knowledge about local seed varieties would be lost. This means that HIV/AIDS constitutes a severe threat to agrobiodiversity. At the request of four communities in Tanzania, several local seed fairs were organized to enable farmers to share and exchange their local knowledge and local seed varieties.

A study in Swaziland looked at the relation between micronutrient intake and HIV/AIDS to establish an inventory of the indigenous foods found in the Manzini region. It also documented the methods of preparation for human consumption and medicinal purposes, according to preference by age, gender and socioeconomic status and farming practices. The study focussed on the issue of food insecurity, as underutilization of indigenous foods contributes to the problem. The seasonal availability of crops was examined so that the periods when specific foods are unavailable were easily identified. This information, also utilized as material for radio programmes and community workshops, was important for the government as well as international agencies planning intervention programmes (Hlanze, Gama and Mondalane, 2005).

3.3 Communication

Communication was a component strategy of all project activities seeking to increase the visibility of men and women's knowledge among communities, development workers and policymakers. Communication at the rural community level was conducted through participatory research processes, encouraging dialogue, feedback to communities and follow-up action that further enhanced learning, and empowerment. The project also promoted communications at the intermediate and policy levels.

- 787 researchers, policymakers and development workers participated in workshops and seminars organized to raise awareness and facilitate discussion of the issues. Several small workshops focussed on exploring the issues of farmers' rights and intellectual property rights. Through these workshops, the project fostered discussion of local knowledge and its link to biodiversity conservation and food security in each of the project countries.
- A wide range of informative material was developed and disseminated. In total, 20 short case studies, 33 research reports and two videos were disseminated to project partners through training workshops, seminars, and the LinKS project mailing list. The project also supported agricultural fairs, contributed to national television and radio programmes, national newspapers and specialist magazines. A website was set up *(www.fao.org/sd/links/gebio.htm)* to disseminate output and provide useful resources and links to information sources.

LinKS collaborated with the World Bank Indigenous Knowledge (IK) Programme to support a government-led effort in Tanzania to develop a national strategy for IK. As an important follow-up to the implementation of this strategy, a trust fund for local knowledge was established in Tanzania for mainstreaming local knowledge at the national level. Moreover, this trust fund aimed to ensure the sustainability of the project's efforts in Tanzania on the long term. The trust acted as a platform for advising the government on LinKS issues in the country, to creating a forum for advocating, promoting, protecting and networking LinKS to ensure its continuous use and sustainability for social-economic development. The trust, which is a non-governmental and non-profit-making organization, was prompted by the need to make LinKS issues visible in national policies and strategies at different levels. The trustees, from eleven different institutions, gave a multidisciplinary nature to the process and offered a good platform for exchanging experiences, and sharing ideas and information on LinKS management issues (Zangari, 2005).

Mozambique and Swaziland, also project countries, expressed interest in a similar process and have established informal networks of different partner institutions that have a specific interest in local knowledge.

3.4 Project Challenges

LinKS was a complex project in terms of its thematic focus, the scope of its activities, the number of countries involved (four) and project management. It not only dealt with the three main issues of gender, local knowledge and agrobiodiversity but also with the linkages between these. Inherent to the thematic focus was the emphasis on gender-sensitive participatory approaches, perceived as the best/only way to develop an understanding of local knowledge and gender issues. Further, the project was implemented in a participatory manner, at least as far as FAO administrative regulations would permit. The participatory management style, together with a holistic approach, was a new and innovative approach for FAO in project implementation. This complexity—both conceptually and logistically—posed numerous problems to those involved in the project. The following section highlights the main challenges and the solutions developed to achieve the project's objectives.

Project Concepts

Each of the three main LinKS themes was a challenge in itself. Over thirty years of research on gender issues point to the difficulties of addressing the gender approach as methodological analytical tool. There are different interpretations, complex theoretical frameworks and several analytical points of reference. Attempts to address local knowledge and agrobiodiversity are similarly complex. For example, 'agrobiodiversity' was perceived by some partners as a new buzzword without a real understanding of its meaning or how to deal with it. Going beyond these individual challenges, LinKS tried to highlight, from the perspective of sustainable livelihoods, how these three themes are interlinked and how they influence each other. The aim of LinKS was to convince partners that only a holistic approach could provide an in-depth understanding and serve as a tool for strengthening food security and sustainable agrobiodiversity management.

The linkages between gender, local knowledge systems and agrobiodiversity management for food security cover a large research area that involves a wide range of cross-cutting issues. These need to be looked at from a holistic and systemic perspective. Only through an interdisciplinary approach, and by integrating different, complementary, disciplines, can a detailed understanding of the complexity be developed. Therefore, research activities need to be designed in a process-oriented way to include the active involvement of all disciplines concerned from planning to implementation, to the analysis and interpretation to ensure a critical reflection of the outcomes FAO, 2003). Such a multidisciplinary manner was extremely challenging, as ministries, universities and most NGOs traditionally work with a sectoral approach.

Attempting to strike the right balance between the three themes, the project experienced difficulties with:

i) Project partners having problems in conceptualizing more than one theme simultaneously;

ii) Project partners having difficulties in establishing clear linkages between the themes;

iii) There was a tendency for 'the concept to fade away' and had to be refreshed from time to time; and

iv) Concepts not always understood accurately (i.e., gender/power; participatory methodologies).

As research progressed, it became clear that both national and international partners experienced difficulties in incorporating gender in a comprehensive way while integrating local knowledge and agrobiodiversity. Research reports and seminar papers reflect some of the difficulties faced by partners in trying to grasp the three themes simultaneously. Some placed more emphasis on local knowledge; paying lip service to agrobiodiversity, while others incorporated more of a gender perspective. In terms of 'gender', many reports showed significant oversights because:

i) Gender was approached in an inconsistent manner, presenting some of the findings disaggregated by sex or analysed along gender lines, or in a gender neutral manner;

ii) Focus was on local knowledge or agrobiodiversity, with little reference to gender; and

iii) 'Gender' was interpreted as 'women' and 'women's knowledge', with little or no comparative data on men or other socioeconomic aspects.

Even the international research institutions involved in the project to provide technical backstopping to the national research teams were often unable to deal adequately with these complexities.

So, how did the project team deal with this conceptual challenge? First of all, efforts were made to clarify the concepts as much as possible. To ensure common understanding, a clear definition was developed for each conceptual term, and a strategy was drafted for each of the three core activity areas: research, capacity-building, and communication and advocacy. The individual strategies were then compiled into one overall project plan. In addition, research guidelines were developed with the support of Noragric.

These measures, however, did not really help the partner institutions and research teams in carrying out the research. It became obvious in both field work and data analysis that despite intensive training and technical backstopping throughout the research period, the application of concepts and approaches was not clear. Pre-field training was offered to ensure that researchers were able to document local knowledge in such a way that was also beneficiary to the local communities – the proprietors of knowledge. In addition, during intervals between field work, time and technical support were allocated to data analysis and to a careful and rigorous reflection of the findings.

Initially, training workshops focussed on the application of gender-sensitive participatory tools within the context of gender, local knowledge and agrobiodiversity. It was assumed that this would also bring a sound understanding of the concepts and their linkages. When it became clear that this was not sufficient, a training manual was developed to address the tools and to clarify the concepts and their linkages. Both methodological and the more conceptual training workshops were complementary.

Participation for all?

The original operational document proposed that the LinKS project be developed with stakeholders in a participatory manner to ensure long-term sustainability. All stakeholders, together with the project team, would be involved in developing and shaping the scope and activities of LinKS project in a participatory manner. At a first glance this did not seem to be an impossible task. However, taking FAO's administrative procedures and its perception of participation into consideration, it turned out to be quite a challenge.

After the first phase of the project, it was clear that there was need for a reassessment of how much participation was feasible, given the various factors inherent to FAO (i.e, a top-heavy and procedure-encumbered institution) that hampered the participatory process (e.g., bureaucracy, hierarchical structures, non-participatory 'cultural' values, etc.). Moreover, the project was totally managed from FAO headquarters, which added another dimension to the problems.

The section below shows how the project management tried to respond to these challenges.

Participatory Project Management

In an attempt to mitigate the participatory 'limitations' posed by the existing institutional framework, LinKS set up a special project structure. National coordination teams with managerial responsibility for project activities were established in each project country, and these were in close contact with the project team at FAO headquarters, who had overall responsibility. As much as possible, this responsibility was delegated to the national teams. For instance, in Tanzania a technical advisory team was created to provide additional technical support to the national team. National team offices were established within the hosting institutions, rather than within the FAO representation. Thus, a much closer collaboration with partner institutions was possible: this was an important element to assure the integration and continuation of LinKS activities in the long-term.

These partner institutions formed informal networks. They met regularly to exchange experiences on LinKS issues and searched for ways to disseminate and mainstream project output and that of their own activities. LinKS staff helped with these participatory networks and strengthened the interface between civil society and government agencies. The development of such a horizontal structure, where all member institutions had the same rights and possibilities to work for the advocacy of local knowledge, was an interesting example of a participatory bottom-up approach. In Tanzania, the network went a step further and created a national trust fund on local knowledge.

Compared to usual FAO projects, the structure of the project and the communication channels were simpler, more flexible and less hierarchical. However, the fact that the project operated in a slightly different way than the conventional FAO project provoked some confusion with FAO colleagues, often hampering project implementation. For example, delays in payments postponed the start of research activities or signing of consultants' contracts. This was a considerable challenge for a participatory project working with farmers who depended on the seasons. A payment late 'only' three or four weeks could easily lead to a half-a-year delay in research activities because certain seasonal activities could not be carried out as planned. In addition, research team members were usually affiliated with different partner institutions, each with their own responsibilities and commitments, which needed attention as well. Such recurrent delays meant that the participatory processes were often interrupted and momentum lost both for the research teams and the communities involved.

3.5 Promoting Participatory Research

The journey had been long between the project's starting point—when the rural community had been 'allowed to participate' in the research study—to the final stage when a research team member pointed out that, 'Farmers are the real specialists! They have their own choices'. In the early stages, the LinKS trainers were often confronted with a 'we know it all' attitude. However, during field work it became evident that most of the workshop participants could never have the opportunity to apply participatory or gender-sensitive tools in a real-life situation. Also, it became clear during the different research studies that the simple application of participatory tools did not go far enough. Anecdotal reports were presented by the research teams, underlining the specific local character of local knowledge and practices. Partner institutions lacked an in-depth understanding of the linkages between local knowledge, gender and agrobiodiversity for food security.

Too often, participatory tools may be considered as 'simple' by formally-trained scientists and researchers. However, the notion of 'simple is easy' is clearly not true for participatory approaches. Experience clearly indicates that internalizing and adopting participatory approaches is a long and iterative process that needs time and commitment from all involved. Over twenty years of global literature on participatory learning approaches highlight the fact that people need intensive guidance and in-depth training both in the uses of participatory tools and in working with communities in ways that do not raise their expectations needlessly. As one team member stated, 'One training session of two weeks does not change people's attitudes that much' nor can it fully provide them with the skills for applying participatory tools and techniques.

A two-week training course cannot fully equip participants to incorporate gender issues and participatory approaches in their work. Iterative approaches to training were much more effective in the long run, allowing researchers and extensionists sufficient time to work with the communities, adapting and revising approaches before trying them again. Experience in LinKS suggests that training prior to a research activity is important but not enough. Over and over, participants indicated the need for post-workshop follow-up monitoring and mentoring to assess the problems faced in attempts to implement what the people had learned at the workshops. LinKS tried to address this through intensive technical support throughout the research process, from research design, data collection and analysis, to interpretation and presentation.

Research reports were shared with the local communities and stakeholders for feedback before being finalized. Such feedback sessions were also important to identify follow-up action with the local communities and stakeholders to ensure that they benefited from the studies. For example, when the seed-system studies in Tanzania identified the need for better access and sharing of local varieties, local seed fairs were organized, giving farmers opportunity to meet, share experiences and exchange their own local variety seeds.

Considering the numerous partners involved in the project, LinKS made a special effort to encompass a wider range of stakeholders in the research with participatory action. After overcoming some initial hesitation, the research teams adapted a simplified version of the PAR (Participatory Action Research) approach. Over time the research teams organized the study in repeated cycles based on methods of reflection-planning-acting-observing. Each time, different stakeholders were involved.

During each round, the research questions and tools were revisited, refined and rendered more focused. Communication became more 'intimate' and barriers reduced, once participants were more increasingly involved. In Tanzania, during the first cycle of the study on traditional seed systems, the participants' overall impression was that a lot of local seed varieties had been lost in the area under study. During the second cycle, however, because of a more focused approach, the research team members identified very knowledgeable farmers and concluded that local seed varieties had not been lost after all. They were still being planted by knowledgeable farmers, but on a very small scale.

3.6 Beneficiaries

Another challenging aspect of the project was to identify the actual beneficiaries and to determine how each could in fact benefit from the project. The original project document outlines the following beneficiaries (Box 2).

The experience from phase one made it clear that, given the timeframe, resources, institutional set-up and scope, it was impossible to reach all the initial beneficiaries. The project did not have the capacity either to work directly with farmers or to focus very much on policymakers. While the project document had anticipated early on quite a few policy and advocacy activities, achieving them was unrealistic. Therefore, during the second phase, LinKS tried to meet the needs of the farmers through intermediaries, i.e., institutions and individuals working with farmers. These mid-level development workers and researchers were

Box 2: Anticipated Beneficiaries of the LinKS Project

Rural men and women: The project involves rural men and women in participatory research and action-oriented activities that will provide them with opportunities to share information, dialogue among themselves, and share experiences. Moreover, an additional means of benefiting this group is to influence the thinking and the approaches used by researchers, government agencies, NGOs and policymakers so that their interactions with rural people are based on respect for and appreciation of their knowledge, needs and perspectives.

Researchers in research institutions and faculty in universities and training colleges: This group benefits from the training activities, which are designed to enhance skills in using gender analysis and participatory research approaches, as well as from other learning opportunities such as workshops, seminars, exchange visits and the dissemination of methods manuals and teaching material. Some researchers will benefit directly from small grants provided by the project to support research activities, and the opportunities associated with this research to gain greater skills in conducting participatory research with rural communities. Universities at large will benefit from the support for curriculum development, opportunities to debate the issues and undertake carry out participatory research.

Mid-level development workers: including staff in NGOs and government agencies or projects that are working with rural communities. This group benefits from the training activities. Some NGOs and government agencies are also directly benefiting from the small grants provided by the project to support their on-going activities.

Policymakers: This includes policymakers in civil society and government, technical units (such as the seed units) of the ministry of agriculture in each country and other key policymakers in the government sector. This group benefits from the opportunities provided by the project to discuss and debate issues and by the information generated or disseminated by the project that can contribute to informed development of national strategies and policies.

considered as an important catalyst for spin-off effects to the academia, government and NGOs. It was believed that the knowledge and experience gained from involvement in the LinKS project would spill over into other areas, such as agricultural extension, natural resources management, and advocacy for local communities, etc. Decision-makers were targeted mainly through participation in awareness-raising workshops and the provision of various information materials.

4. Summary of Main Findings

The LinKS project did not emerge in a vacuum. Instead, it continually attempted to build on existing activities and initiatives, trying to reinforce and strengthen partner institutions in issues on gender, sustainable management of agrobiodiversity and local knowledge. Thus LinKS reinforced existing trends and tendencies. An increasing interest, particularly in Tanzania, in the three themes was noted. The project's efforts to strengthen local knowledge were reinforced by the World Bank's indigenous knowledge programme and by several national institutions working together in a complimentary manner. Several countries—for example, Uganda and Kenya—showed interest in local knowledge research.

Capacity-building for participatory approaches, gender analysis and local knowledge was a time-consuming exercise. The LinKS experience clearly indicated that it was not enough to provide people with one or two training workshops and then expect them to apply what they had learned. People needed time and opportunity to apply newly acquired techniques in day-to-day working situations. The big challenge was providing sufficient time and opportunity to make sure that people understand the approaches and tools, to apply them and therefore to change their thinking. This was very time consuming. Furthermore, this amount of time had never been included in any of the work plans or budgets.

Most international organizations claim to work in a participatory way and to apply gender analysis and tools. Is this lip-service? The answer is no. In fact, most of the workshop attendants had already participated in several similar training sessions and felt that they knew it all beforehand. But once in the field, they were unable to use the approaches and tools properly and coherently. A closer look at the issue revealed that little had been achieved by the quick and often limited inputs provided by donor organizations training workshops.

One gap became profoundly evident while working with the teams in the field: many researchers were unable to analyse socioeconomic data and to report research results in a coherent and well explained manner. Also, the combination of qualitative and quantitative data, their analysis and presentation created a challenge. Research reports frequently consisted only of tables or anecdotal stories. Thus, the need for capacity-building and for developing appropriate training material was great.

But some interesting research results were observed in relation to seeds, plant genetic and animal genetic resources. Studies in Swaziland, Mozambique and Tanzania highlighted interesting findings with regard to the link between local knowledge and agrobiodiversity, particularly the effects of HIV/AIDS on seeds management. The relation between the epidemic—affecting primarily women—and the consequent loss of female crop knowledge of seed varieties had previously been unknown. Another interesting point was the extremely limited exchange of information between husband and wife, leading again to a loss of knowledge and of agrobiodiversity. These studies emphasized the importance of underutilized crops—not cash crops that are used for marketing, but food crops for survival.

Food crops were still vital for the rural population. In Tanzania, for example, farmers did not depend on the formal system to any extent. During the first round of research, most of the local diversity in seed variety appeared to have been completely lost. A deeper analysis, however, showed that most were still available but in a very small scale, with only a few knowledgeable farmers. On the other hand, improved varieties, where available, were often not affordable to farmers, as these were sold in very large quantities. Research extension staff's knowledge of local seed varieties was limited and therefore formal and informal seed systems really did work in parallel. There was also a distinction between the crops farmed by women (food crops) and by men (cash crops), but this appeared to be quite flexible and dependent on market fluctuations.

With regard to animal genetic resources in particular, an ongoing study on livestock in Tanzania looks at the Masaai society to examine local knowledge, and the roles and responsibilities of women in connection with animal genetic resources. According to preliminary conclusions, the local knowledge of the Maasai of Simanjiro is alive and dynamic, and widespread among all members of the Maasai society. The extent to which local knowledge is maintained and practised differs according to age and gender. Knowledge is passed along vertical lines from older members of the society to the younger groups through instruction and initiation. But information is also exchanged horizontally through interaction with peers, through personal contact, and through contact with the outside world (travel, markets). The research team gained a better understanding of the concept of local knowledge and its relation to project objectives. The link between local knowledge and community preferences, and the criteria for breeding and selection were well documented. The team gained further insight into the approaches and methods of conducting social research, and in understanding that a difference exists between informal and participatory research.

4.1 The Way Forward

FAO has long recognized the strong linkages between the different knowledge of men and women, their skills and biodiversity so critical to agricultural production and food security, and has supported various initiatives in Asia, Africa, and Latin America. Given FAO's mandate within the UN system for food and agriculture, it has an important role to play also at the level of international policy. Much support was provided through the 1990s to make sure that gender remained on the international agenda, particularly in arenas such as the Commission on Genetic Resources for Food and Agriculture (CGRFA).

Based on FAO's experience with the LinKS project, the following points need further attention.

As mentioned in the introduction, the International Treaty on Plant Genetic Resources for Food and Agriculture is an important step in bringing together governments, farmers and plant breeders as it offers a multilateral framework for accessing genetic resources and sharing benefits. So far, negotiations and discussions have taken place at an international level. However, for treaty implementation, the signatory countries need to develop tools and mechanisms for the national level. Furthermore the treaty does not focus on the gendered nature of local knowledge. Additional action is required to ensure that distinctions are made, where appropriate, between the different knowledge bases and access of resources of women and men.

The trust fund, created in Tanzania and focusing on local knowledge and agrobiodiversity, could function as a national support instrument for the implementation of the Treaty. It could serve as a platform for experiences on local knowledge and agrobiodiversity. Moreover, it could help to clarify the process on issues surrounding farmers' rights, and to define who, in Tanzania, are the 'farmers'. The enormous effort of bringing the Treaty into force and all the mechanisms and instruments that still need to be developed will be successful only if policy institutions recognize that men and women farmers play an important and crucial role in the management and conservation of plant genetic resources. Furthermore, men and women farmers need to participate actively in the decision-making processes and make use of their right to share equitably the benefits arising from the utilization of plant genetic resources.

More work is needed to understand the institutions, associated constraints and incentives influencing relevant actors in their use of the LinKS concepts in their daily work. For example, the Plant Breeders Rights Act in Tanzania provides the incentive for breeders and others to develop and release new varieties (a percentage of the royalties from the sale of seed should go to the breeder and organization responsible for release), but there are few incentives for researchers and extension workers to assist men and women farmers to better manage their own seed (which comprises the vast majority of the seed planted in Tanzania).

Increased productivity, economic growth and agricultural productivity are important elements in poverty reduction. The diverse and complex agroecological environment of Sub-Saharan Africa will direct future efforts on more localized solutions, instead of an 'Asian-type green revolution'. This means that future activities will have to build much more on local knowledge and agrobiodiversity with a clear understanding of gender implication.

(Yianna Lambrou and Regina Laub, Gender and Development Division, Food and Agriculture Organization of the United Nations, Rome. They can be reached at yianna.lambrou@fao.org, Regina.Laub@fao.org respectively.)

ANNEX I

Short Overview of the LinKS Project Phase I and II (1997-2005)

Project Countries

Mozambique, Swaziland, Tanzania and Zimbabwe (until May 2002)

Budget

US$3,813,953

Development Goal

Enhance rural people's food security and promote sustainable management of agrobiodiversityby strengthening the capacity of institutions in the agricultural sector to apply approaches that recognize men and women farmers' knowledge in their programmes and policies.

Immediate Objectives

- Enhance the ability of researchers and development workers from key partner organizations to apply an understanding of gender, local knowledge, biodiversity and food security in their work by providing them with diverse learning opportunities as well as skills enhancement in gender-sensitive and participatory approaches.
- Increase the visibility of men and women's knowledge about the use and management of agrobiodiversity among key development workers and decision-makers by supporting documentation of good practices, research and communication.
- Enable partner organizations and policymakers to network, develop guidelines and strategies, and take action to promote a greater recognition of rural people's knowledge, needs and perspectives by providing financial and technical support for partner's initiatives at all levels.

Strategy

- Basic strategy of building on and 'adding value' to the ongoing activities of partner organizations.

- Enabling local initiatives for mainstreaming and institutionalization.
- Decentralized decision-making processes.

Beneficiaries

Men and women farmers as custodians of knowledge; development workers, researchers and staff from institutions in the agriculture and environment sectors.

Structure

- National country teams consisting of a national coordinator, a project assistant and a project driver facilitate activities in each country.
- 'Hosting institutions' provide housing and support services. and
- The Gender in Development Service at FAO provides overall management and coordination.

ANNEX II

Acronyms

CBD	Convention on Biological Diversity.
CGIAR	Consultative Group on International Agricultural Research.
CGRFA	FAO's Commission on Genetic Resources for Food and Agriculture.
IPGRI	International Plant Genetic Resources Institute.
IPR	intellectual property rights.
ITPGR	International Treaty on Plant Genetic Resources.
ITPGRFA	International Treaty on Plant Genetic Resources for Food and Agriculture.
LinKS	A regional FAO project on local indigenous knowledge, gender and biodiversity in Mozambique, Swaziland, Zimbabwe and Tanzania.
MDGs	Millennium Development Goals.
NGOs	Nongovernment organizations.
Noragric	The Department of International and Development Studies at the Norwegian University of Life Sciences.
PAR	Participatory action research.
PGRFA	Multilateral System of the International Treaty on Plant Genetic Resources for Food and Agriculture.
PVP	Plant variety protection.
SOFI	FAO's annual report, The State of Food Insecurity in the World.
TPGR	Treaty on Plant Genetic Resources.
WFS	World Food Summit.
WSSD	World Summit on Sustainable Development.

Bibliography

Adams W M, R Aveling D Brockington, B Dickson, J Elliott, J Hutton, D Roe, B Vira and W Wolmer (2004), 'Biodiversity Conservation and the Eradication of Poverty'. *Science,* 306 (5699): 1146-49.

Altieri M (2004), 'Linking Ecologists and Traditional Farmers in the Search for Sustainable Agriculture'. *Frontiers in Ecology and the Environment,* 2 (1): 35-42.

Andersen R (2003), 'FAO and the Management of Plant Genetic Resources', in O. Schram Stokke and Ø. B. Thommessen (eds), *Yearbook of International Cooperation on Environment and Development 2003/2004.* London: Earthscan Publications.

Barany M, A L Hammett, A Sene, and B Amichev (2001), 'Nontimber Forest Benefits and HIV/AIDS in Sub-Saharan Africa'. *Journal of Forestry,* 99 (12): 36-41.

Barber J (2003), 'Production, Consumption and the World Summit on Sustainable Development'. *Environment, Development and Sustainability,* 5 (1-2): 63-93.

Beintema N M, and G–J Stads (2004), 'Sub-Saharan African Agricultural Research: Recent Investment Trends'. *Outlook on Agriculture,* 33 (4): 239-46.

Borowiak C (2004), 'Farmers' Rights: Intellectual Property Regimes and the Struggle over Seeds'. *Politics and Society,* 32 (4): 511-43.

Brush S B (2005), 'Farmers' Rights and Protection of Traditional Agricultural Knowledge'. CAPRi Working Paper No. 36. Washington, DC: International Food Policy Research Institute (IFPRI).

Calderón J (2005), 'Lessons from an Activist Intellectual: Participatory Research, Teaching, and Learning For Social Change'. *Latin American Perspectives,* 31 (1): 81-94.

Cooper H D (2002), 'The International Treaty on Plant Genetic Resources for Food and Agriculture'. *Review of European Community and International Environment Law,* 11 (1): 1-16.

Deda P and R Rubian (2004), 'Women and Biodiversity: The Long Journey from Users to Policy Makers'. *Natural Resources Forum,* 28 (3): 201-4.

Dovie D B K (2002), 'Towards Rio + 10 – Trend of Environmentalism and Implications for Sustainable Livelihoods in the 21st Century, the Context of Southern African Region'. *Environment, Development and Sustainability,* 4 (1): 51-67.

Dudgeon R C and F Berkes (2003), 'Local Understandings of the Land: Traditional Ecological Knowledge and Indigenous Knowledge', in H Selin (ed.), *Nature Across Cultures.* Dordrecht: Kluwer, 75-96.

Food and Agriculture Organization (FAO) (2000). 'Benefits and Risks of Sharing Local Knowledge'. Summary Report with Draft Research Guidelines'. LinKS Technical Report No. 3. Rome: FAO.

Food and Agriculture Organization (FAO) (2003), 'Draft Research Strategy for the LinKS' project'. Geneva: FAO. Available at: *www.fao.org/sd/LINKS/documents_ download/ LinKSprojectstrategy.pdf*

Food and Agriculture Organization (FAO) (2004), *The State of Food Insecurity in the World: Monitoring progress towards the World Food Summit and Millennium Development.* Rome: FAO.

Food and Agriculture Organization (FAO) (2004X), 7th LinKS Training Workshop on Gender-Sensitive Participatory Approaches to Research and Extension on Local Knowledge in Agrobiodiversity Management for Food Security, 12-23 July, Kiaha, Tanzania.

Food and Agriculture Organization (FAO) (2005), 'Gender Sensitive Participatory Approaches to Research and Extension Staff on Local Knowledge in Agrobiodiversity Management for Food Security'. Documentation of the 9th LinKS Training Workshop, 3-14 January, in Tanga, Tanzania.

Fowler C (2004). 'Accessing Genetic Resources: International Law Establishes Multilateral System'. *Genetic Resources and Crop Evolution,* 51 (6): 609-20.

Gari J A (2002), *Agrobiodiversity, Food Security and HIV/AIDS Mitigation in Sub-Saharan Africa,* SD Dimensions. Rome: FAO.

Gillespie S, L Haddad and R Jackson (2001), 'HIV/AIDS, Food and Nutrition Security: Impacts and Actions'. Paper presented at the 28th Session of the ACC/SCN Symposium on Nutrition and HIV/AIDS, 1 May.

Gladis T (2003), 'Agrobiodiversity with Emphasis on Plant Genetic Resources. *Naturwissenschaften,* 90 (6): 241-50.

Gollin D and R Evenson (2003), 'Valuing Animal Genetic Resources: Lessons from Plant Genetic Resources'. *Ecological Economics,*. 45 (3): 353-63.

Goma H C, K Rahim, G Nangendo, J Riley and A Stein (2001), 'Participatory Studies for Agro-Ecosystem Evaluation'. *Agriculture, Ecosystems and Environment,* 87: 179-90.

Groombridge B and M D Jenkins (2002), *World Atlas of Biodiversity: Earth's Living Resources in the 21st Century.* Berkeley: University of California Press.

Grown C, G Rao Gupta and A Kes (2005), *UN Millennium Project 2005. Taking Action: Achieving Gender Equality and Empowering Women.* Task Force on Education and Gender Equality. London: UNDP/Earthscan for the Task Force on Education and Gender Equality.

Hlanze Z, T Gama and S Mondalane (2005), 'Impact of HIV/AIDS on Drought and Local Knowledge Systems for Agrobiodiversity and Food Security'. Indigenous Crops Report 25/03/ 05. Geneva: FAO. Available at: *www.fao.org/sd/LINKS/documents_download/HIV-AIDS.pdf*

Howard P (ed.) (2003), *Women and Plants: Gender Relations in Biodiversity Management and Conservation.* London: Zed Books.

Jayne T S, M Villarreal, P Pingali and G Hemrich (2004), 'Interactions between the Agricultural Sector and the HIV/AIDS Pandemic: Implications for Agricultural Policy'. ESA Working Paper No. 04-06. Rome: Agricultural and Development Economics Division, FAO.

Johnson N, N Lilja, J A Ashby and J A Garcia (2004), 'The Practice of Participatory Research and Gender Analysis in *Natural Resource Management*'. Natural Resources Forum, 28 (3): 189-200.

Kameri-Mbote P (2004), 'The Impact of International Treaties on Land and Resource Rights', in Munyaradzi Saruchera (ed.), *Programme for Land and Agrarian Studies.* Geneva: International Environmental Law Research Centre.

Kengni E, C M F Mbofung, M F Tchouanguep and Z Tchoundjeu (2004), 'The Nutritional Role of Indigenous Foods in Mitigating the HIV/AIDS Crisis in West and Central Africa'. *International Forestry Review,* 6 (2): 149-60.

Laird S A (ed.) (2000), *Biodiversity and Traditional Knowledge: Equitable Partnerships in Practice.* London: Earthscan.

Lambrou Y and R Laub (2004), 'Gender Perspectives on the Conventions on Biodiversity, Climate Change and Desertification'. Rome: FAO.

Louwaars N (2000), 'Seed Regulations and Local Seed Systems'. Biotechnology and *Development Monitor,* 42: 12-4.

Maier J (ed.) (2002), 'Livestock Diversity: Keepers' Rights, Shared Benefits and Pro-Poor Policies'. Documentation of a Workshop with NGOs, Herders, Scientists, and FAO'. Bonn: German NGO Forum on Environment and Development.

McMichael A J (2004), 'Environmental and Social Influences on Emerging Infectious Diseases: Past, Present and Future'. *Philosophical Transactions of the Royal Society of London Series B Biological Sciences,* 359 (1447): 1049-58.

McNeely J A and S J Scherr (2001), 'Common Ground, Common Future: How Ecoagriculture can Help Feed the World and Save Wild Biodiversity'. IUCN-The World Conservation Union Report No. 5/01. Gland: IUCN and Future Harvest.

Mkuchu M (2006), 'A Study of Local Knowledge in Relation to Management of Agrobiodiversity and Food Security in Southern Highlands, Tanzania'. Geneva: FAO. Publication under preparation.

OECD DAC (Development Assistance Committee) (2001) 'Poverty-Environment-Gender Linkages'. OECD DAC Journal 2, No. 4. Available at *www.oecd.org/dataoecd/47/46/1960506.pdf.*

Oldham P (2002), 'Negotiating Diversity: A Field Guide to the Convention on Biological Diversity'. Lancaster: CESAGEN and Lancaster University.

Oxfam International (2005). 'Paying the Price: Why Rich Countries Must Invest Now in a War on Poverty'. Oxford: Oxfam International. Available at: *www.oxfam.org.uk/what_we_do/issues/debt_aid/mdgs_price.htm*.

Painter G R (2004), 'Gender, the Millennium Goals, and Human Rights in the Context of the 2005 Review Processes'. Report for the Gender and Development Network. London.

Piana G and Y Lambrou (2005), 'Gender, the Missing Component in the Response to Climate Change'. Rome: FAO.

Pound B, S Snapp, C McDougall and A Braun (eds) (2003), *Managing Natural Resources for Sustainable Livelihoods: Uniting Science and Participation: Using Participation to Improve Natural Resource Management.* Ottawa: International Development Research Centre (IDRC).

Pretty J and D Smith (2004), 'Social Capital in Biodiversity Conservation and Management'. *Conservation Biology,* 18 (3): 631-8.

Ramírez R and W Quarry (2004), 'Communication for Development: A Medium for Innovation in Natural Resources Management'. Ottawa: International Development Research Centre and Rome: FAO.

Roe D (ed.) (2004), 'The Millennium Development Goals and Conservation: Managing Nature's Wealth for Society's Health'. London: International Institute for Environment and Development.

Ruiz M (2004), 'Access to Genetic Resources, Intellectual Property Rights and Biodiversity: Processes and Synergies'. Gland: IUCN-The World Conservation Union.

Sanchez P, M S Swaminathan, P Dobie and N Yuksel (2005), UN Millennium Project 2005. *Halving Hunger: It Can Be Done.* Task Force on Hunger. New York: UNDP and London: Earthscan.

Simon D, D McGregor, K Nsiah-Gyabaah and D Thompson (2003), "Poverty elimination, North-South research collaboration, and the Politics of participatory Development'. *Development in Practice,* 13 (1): 40-56.

Spangenberg J H (2002), 'Institutional Sustainability Indicators: An Analysis of the Institutions in Agenda 21 and a Draft set of Indicators for Monitoring their Effectivity'. *Sustainable Development,* 10: 103-15.

Srinivasan C S (2003), 'Exploring the Feasibility of Farmers' Rights'. *Development Policy Review,* 21 (4): 419-47.

Thrupp L A (2000), 'Linking Agricultural Biodiversity and Food Security: The Valuable Role of Agrobiodiversity for Sustainable Agriculture'. *International Affairs,* 76 (2): 265-81.

Tripp R (2000), 'Strategies for Seed System Development in Sub-Saharan Africa: A Study of Kenya, Malawi, Zambia, and Zimbabwe'. ICRISAT Working Paper Series 2. Bulawayo: International Crops Research Institute for the Semi-Arid Tropics (ICRISAT).

Tripp, R. (2002). 'Sowing the Seeds, Strengthening the Roots – Seed System Development in Sub-Saharan Africa', *Strategies for Seed System Development in Sub-Saharan Africa.* London: ODI and Patancheru: International Crops Research Institute for the Semi-Arid Tropics (ICRISAT).

UNDP (2003). *Human Development Report: Millennium Development Goals: A Compact Among Nations to End Human Poverty.* New York: Oxford University Press.

UNESCO (2002), 'Indigenous Knowledge'. *International Social Science Journal,* 173. Special Issue on Indigenous Knowledge. Available at: portal.unesco.org/shs/en/ev.php-url_id=3136&url_do=do_topic&url_section=201.html

International Council for Science (ISCU) (2002), 'Science, Traditional Knowledge and Sustainable Development'. ISCU Series on Science for Sustainable Development No. 4. Paris: ICSU.

Van Vlaenderen, H. (2004), 'Community Development Research: Merging Communities of Practice'. *Community Development Journal,* 39 (2): 135-43.

Vernooy, R. (2003). *Seeds that Give: Participatory Plant Breeding.* Ottawa: International Development Research Centre (IDRC).

Villalobos R, B Guiselle, M Blanco Lobo and F A Cascante (2004), 'Diversity Makes the Difference: Actions to Guarantee Gender Equity in the Application of the Convention on Biological Diversity'. Gland: The World Conservation Union-IUCN.

Von Braun J, M S Swaminathan and M W Rosegrant (2004), 'Agriculture, Food Security, Nutrition and the Millennium Development Goals'. IFPRI Annual Report Essay. Rome: International Food Policy Research Institute.

Wollny C B A (2003), 'The Need to Conserve Farm Animal Genetic Resources in Africa: Should Policy Makers be Concerned?'. *Ecological Economics,* 45 (3): 341-51.

World Bank Group (2003), *Poverty and Climate Change – Reducing the Vulnerability of the Poor Through Adaptation.* Washington, DC: World Bank.

Zangari A (2005), 'The Potential Importance of a Trust Fund and its Possible Roles'. Final Trust Fund Concept Paper. Available at: *www.fao.org/sd/LINKS/documents download/trustfund.pdf*.

10

Indigenous Knowledge and Biodiversity Conservation and Management in Ghana

Luc Hens

An analysis of a series of biodiversity related areas in Ghana, including ecosystem, water and soil management, farming, fishing and hunting practices and the collection of herbal medicines, shows that Indigenous Knowledge (IK) has the potential to contribute to the conservation of species, genes and ecosystems. Moreover, as these traditional rules are owned by the locals, they might be implemented more rigorously than governmental laws. However, in the contemporary fast changing societies of sub Saharan Africa, IK is less and less applied and at risk of disappearance. In spite of important paradigmatic differences between IKS and modern science, traditional knowledge is at its best when it is matched with scientific approaches. Eminent examples of this can be found in the development of medical drugs that stem from herbal medicines. Similar approaches are needed and possible for biodiversity conservation. They should focus on questions on effectiveness, efficiency and monitoring of situations where IK is brought into biodiversity conservation and management. At the governmental policy level it is

Source: Journal of Human Ecology (Volume 20, No.1, Pages 21-30, 2006).

remarkable that IK has currently no place in biodiversity conservation strategies. Nevertheless there is a high need to involve local communities in such policies. This is illustrated by providing two examples: the urgent need to curb fundamentally Ghana's logging policy of its tropical forest and the need to involve local communities in the management of the Kakum Conservation area.

1. Introduction

Biodiversity loss has been a major concern to mankind, especially during the last quarter of the previous century. This concern culminated in the "Biodiversity Convention" that was opened for signature at the United Nations Conference on Environment and Development (UNCED) in Rio de Janeiro, Brazil, June 1992. Since then different international fora, including the Beijing Conference for Women in 1995, echoed the problems of continuing environmental degradation. In spite of this, ten years after Rio, the World Summit on Sustainable Development (WSSD) that was held in Johannesburg, South Africa, August-September 2002, could only state that in spite of significant efforts, the loss of biodiversity worldwide was continuing at an unpreceeded speed and that a reverse in this ongoing decline should urgently be realized (Hens and Nath, 2003).

What are the causes of this continuing loss of species, genes and ecosystems? They are multiple and complex. However, one of the traditionally important ones was the unique focus on the biological reasons of biodiversity loss (Box 1 lists the common biological causes of biodiversity loss). The collateral of this focus was that if pollution would be halted, habitat loss be stopped and (the effects of) the introduction of non-indigenous species be reversed, biodiversity would be safeguarded. In managerial terms this meant conservation of biodiversity by creating parks and reserves with minimal to hardly any human influence.

This strategy proved to be insufficient and it was realized that it failed to bring the human factor on board. Demographic changes, economic imperatives, educational and cultural transitions and policies implicitly degrading biodiversity were at least as important for the regression of biological systems than the biological

Box 1: Biological Causes of Biodiversity Loss

Extinction of species is part of an evolutionary process. However, during recent times, extinction rates are ten to hundred times higher than during pre-human times (Sinclair, 2000a). The main biological causes for this loss of biodiversity are:

a) *The loss of habitats:* On human disturbance of habitats on a worldwide scale data show the significant impact of human activity on world ecosystems; For example, in Europe only 15% of the continent is classified as "undisturbed", which is the lowest percentage worldwide. loss of tropical forest is the most highly published aspect of this (Sinclair, 2000b). Elsewhere, rivers are impounded, coral reefs destroyed by dynamite, and natural grasslands are ploughed.

b) *The introduction of exotic species:* Many are accidental, as with noxious weeds and insect pests. Others are deliberate. Foxes, rabbits and cats, which came to Australia aboard of European ships, have decimated Australia's indigenous wildlife. In freshwater, the stocking of exotic fish for sport, or (rarely) for food, has caused at least 18 extinctions of fish species in North American rivers. Catastrophic changes in the fish biodiversity of Lake Victoria (East Africa) resulted from the introduction of Nile perch. Eucalyptus, which is indigenous in Australia, has been introduced in many tropical and subtropical regions in the world, where the tree merely behaves as a pest.

c) *Over-harvesting* by (illegal) hunting, and the systematic cutting of wood for heating purposes, or charcoal production, are other reasons for biodiversity loss. The use of medicinal plants might illustrate this point. In the semi-arid rural area of Southern Cochabamba (Bolivia) it was shown that out of 132 inventarised plants the local people use for traditional medicinal purpose, 10 were threatened because of their intensive collection (Urena Hinojosa, 2001).

d) Lesser known causes are due to "knock-on" effects. Species that are co-evolved with another, such as plants with specialised insect pollinators, will go extinct if one of the pair goes extinct; When the last passenger pigeon (Ectopistes migratorius) died in the early 1990s, so also did two of its obligate parasites, two louse species. Moabi (Baillonella toxisperma) used to be a common tree in West Africa. The fruits are eaten, cooking oil is extracted from the seeds (Karite) and the bark is used for medicinal purposes. For its reproduction the plant depends on the elephants. Only these animals swallow and disperse the moabi seeds. The impressive reduction of elephants in countries as Ivory Coast, Ghana and Benin has an important impact on the distribution of the tree.

e) *Homogenisation in agriculture and forestry:* Although an estimated 7,000 plant species have been collected and cultivated for food, only 30 contribute over 90% of the entire global population energy needs. The case of the banana (Musa spp.) is illustrative. Bananas are the fourth most important food source in the tropics after rice wheat and corn. They are cultivated in nearly 120 countries. Farmers use only about 25 edible sterile banana varieties. The number of varieties is diminishing due to the spread of pests and diseases and the deterioration of the resource.

Contd...

Contd...
f) Pollution and global environmental change also threaten the world's biodiversity. All these causes have one element in common: they are induced by human activity. This makes human activity the most important source of the current decline in biodiversity. Therefore, understanding the many aspects of human influences on biodiversity, and their underlying driving forces, is of crucial importance for setting priorities and counteracting the current negative trends.

causes seemed to be (Hens and Boon, 2000). Only, they proved to be much more difficult to be integrated in biodiversity policies.

Next to a more human ecological approach to biodiversity loss, its management is looking for other, new, effective approaches to reverse the current declining trends. In this context attention is given to "Indigenous Knowledge" (IK). This is about the knowledge that women and men, families and communities have developed themselves for centuries and allowed them to live in their environment for often long periods of time.

In a context of contemporarily development, evidence shows that IK can help to solve local problems, that it offers a resource to help grow more and better food, that it adds to maintain healthy lifestyles, that it provides opportunities to share wealth and prevent conflicts (Mkapa, 2004). In this paper the question is addressed whether and how IK can also contribute to biodiversity conservation. The issue is approached in a three-fold way:

- What is the nature of the IK-biodiversity conservation relationship? In this section a number of examples on how IK can contribute to biodiversity conservation are discussed.
- A specific area of academic importance is the tension between IK and science and technology. A description of this tension and a discussion on how to mach these two knowledge systems subject to different paradigms is the substance of the second section.
- The last part is about how to introduce IK elements in bioconservation policy. The original data in this paper relate to Ghana (West Africa). Through research and the organization of seminars (see e.g., Amlalo *et al.*, 1998) this country has shown a continuing and long standing interest and

experience with IK. The data for this paper were gathered during interviews with experts (see acknowledgement) in July 2005.

2. Indigenous Knowledge and Biodiversity Conservation

The potential of IK to contribute to the conservation of biodiversity is often illustrated using a set of positive examples. Table 1 lists 11 of these examples documented in Ghana in six areas that are relevant for biodiversity conservation.

In the Ashanti region of South Western Ghana, trees, which were regarded as housing spirits should not be felled without performing rituals. This custom had a protective effect on trees as odum (Chlorophora excelsa), African mahogany (Khaya ivorensis) and tall palm trees as betene (Elaeis Guineensis) and osese (Funtumia sp.) also shea butter (Butyrospermum parkii) and the Dawadawa (Parkia clappertoniana) trees, in the Northern savannah zone of Ghana are subject to the same traditional protection system (Boaten, 1998).

Animals in a particular habitat may be regarded as sacred and are therefore protected from hunting. In Ghana this applies to the Black and White colobus (Colobus polykomos) and the mona monkey (Cercopithecus mona) in the

Table 1: Examples of Biodiversity Conservation Related Practices in Ghana

Area of Activities	Practices
Ecosystem preservation	Trees that are regarded as housing spirits: odum, African mahogany, tall palm trees.
	Sacred animals: black and white colobus, mona monkey.
	Totem animals and associated species.
	Sacred groves.
Water	Vegetation cannot be cleared along a strip of 30 m at both banks of streams and rivers.
Farming	Traditional, 10 year, bush fallow system.
	Traditional crops as cocoa and vegetables.
Fishing	Days and periods of banned fishing.
Hunting	Periods of banned hunting.
	Do not hunt pregnant females.
Herbal medicines	Use of herbs to prevent and treat (common) diseases in humans, animals and plants.

Boabeng-Fiema wildlife sanctuary of Central Ghana (Ola-Adams, 1997). A similar situation is reported for the bats of Wli in the South Eastern part of the country. The overhanging rocks of the mountains that form the border with Togo, house an impressive colony of large bats, that are said to be conserved by the local community.

In Ghana, almost every traditional ruler, chief or King has a totem. Many wildlife species are regarded as totems due to their historical or socio-cultural significance. Totem animals vary significantly over tribes and clans. They include merely mammals (leopard, elephant, lion, monkey, buffalo) and birds (falcon, raven, parrot).

But also turtles, crocodiles, snakes (python), scorpions, crabs and fishes are totems. Some totem animals as the crested porcupine (Hystrix cristata) and the patas monkey (Cercopithecus patas) are threatened. There is a belief that an intimate relationship exists between the totem animals and the tribe. Therefore, the members do not eat, kill or trap these animals (Conservation International Ghana, 2005).

Sacred groves are pieces of land set aside for spiritual purposes. They are dotted all over Ghana. They range from a few square meters to several hectares. The larger ones often form distinct elements in the landscape. An example is the sacred grove of Sefwi Wiaswo, near the border with Ivory Coast. This hill is a classical example of a virgin tropical forest surrounded by the city of Sefwi Wiaswo and its deforested hinterland. Apart from the collection of medicinal herbs (after the agreement of the elders), and their use as burial ground, the sacred grove areas are untouched. Farming, hunting, burning, tree cutting and firewood gathering are prohibited. The results of a comparative plant analysis (Table 2) showed a richer floristic composition and a more complex ecosystem structure in the sacred grove of Ikire-Ile (Nigeria) than in the surrounding areas (Alabi, 1992).

Traditional farming practices are champions in sustainable land and water management. They involve land rotation and shifting cultivation allowing the land for more than 10 years to restore its natural fertility. In many places in Ghana, this is now replaced by a pseudo-scientific agriculture in which shifting cultivation is still the practice, but the fallow period is between one to two years only. This calls for nitrates and phosphates to boost crop yield, but the majority of the farmers cannot afford to apply the recommended rates of fertilizers. This unfavorable situation in which farmers do not practice anymore traditional

Table 2: Floral Characteristics of Arable Crop Farmland, Tree Crop Farmland and Sacred Grove at Ikire-Ile Nigeria (Alabi, 1992)

Arable		Crop Farm-land	Tree Crop Farmland	Sacred Grove
1)	No. of families	34	46	67
2)	No. of species	58	72	155
3)	Trees	14	30	54
(a)	Mean dbh (cm)	24,31	29,64	23,91
(b)	Mean Tree height (cm)	8,19	9,44	11,95
(c)	Mean Grown depth (m)	4,92	5,99	7,62
(d)	Mean Grownwidth (m)	3,85	5,6	6,86
(e)	Mean Tree spacing	7,07	3,54	3,23
4)	Shrubs	14	13	31
5)	Herbs	27	21	59
6)	Climbers	3	8	12
7)	Rainkieer's Life-Forms		(No. of species)	
(a)	Megaphanerophytes	3	8	20
(b)	Mesophanerophytes	11	22	34
(c)	Microphanerophytes	14	13	30
(d)	Nanophanerophytes	16	12	43
(e)	Chamaephytes	4	1	11
(f)	Hemicryptophytes	7	8	5
8)	Complexity Index	1,70	48,95	230,57

agriculture, nor sound scientific agriculture, leads to low crop yields and accelerated soil erosion of farmlands (Bonsu *et al.*, 2000). The system is under even higher pressure by the selection of crops. Traditional farming systems use a wide variety of species e.g. vegetables or rice. Current insights strive towards a higher yield, provided by non-indigenous varieties, as a rule requiring higher inputs. Because of the cost of these inputs, they cannot be provided by the farmer. All this results often in a downwards spiral of erosion, lower yields, more poverty and a destroyed environment. From a biodiversity conservation point of view, traditional farming practices entailed strong elements of long-term land rotation and conservation of indigenous plants. However, these are threatened by the current widespread practices of trying, but not succeeding, in applying agricultural practices according to scientific standards.

Coastal ethnic groups know days when they do not fish. Tuesdays and Fridays in Ghana are often set aside and people and the ecosystem were expected to rest. In addition, many lagoons have long periods during which no fishing is allowed. This resting period coincides with the period when the fishes lay their eggs. In the Fesu laguna, just West of Cape Coast, the banned period is during the months of May and June. This period is obviously linked to the procreation of the fishes including the youngsters to mature.

The importance of wildlife in the diet of the Ghanaians has been well documented. Grasscutter is a delicacy all over the country and one of the most important culinary honors in the North is serving your guests an antilope. Nevertheless, hunting and trapping wild animals is subject to restriction. Some animals are regarded as sacred and therefore may not be touched, killed or eaten. Traditionally, it was a widespread offence to kill a fertile and definitely a pregnant animal. The killing is restricted to male and older animals. Certain species could not be hunted during certain seasons (such as the breeding season). In this way the communities were able to ensure continued population growth of their wildlife resources.

All over sub-Saharan Africa, indigenous plants are used in preventing and curing diseases in plants, animals and humans. This equally applies to Ghana. Although the country has established a countrywide network of clinics and other Western health services, the development of the health insurance is still in its infancy. This explains to a large extent the current problems of access to Western medical drugs. In the rural areas, it is a common practice to perform abdominal surgery under local anesthesia, to reduce the cost of the anesthetics. Still today, herbal medicines are more affordable, accessible and used by the majority of the Ghanaians. They are used both for relatively banal indications as digestives or related gastro-intestinal problems, and also to treat malaria, wound care in diabetes patients and even HIV/AIDS. In this context the Mampong-Akwapim based Centre for Scientific Research into Plant Medicine (CSRPM) has a pivotal role. It was most instrumental in the publication of the Ghana Herbal Pharmacopoeia (Policy Research and Planning Institute, 1992). Its herbarium with specimens of reported medicinal plants in Ghana, totals over seven hundred and fifty species. In 1994 it used 64149.50 kg of dry weight of plants and parts of plants to produce herbal medicines for the centre's clinic only. In 1995 the estimated value was 141923.75 kg or an increase of more than 200% (Sam, 1998). The conservation of these plants, offers more and more a defeat for biodiversity conservation.

As shown in Table 3, the application and maintenance of these traditional rules is mainly based on taboos that most often have a traditional religious background. More seldom, customs rooted in tradition, as the landownership issue, drive the traditional approaches. In some cases the rules are maintained by physical sanctions. Not respecting the fishing bans in the Fesu lagona, is punished by the chief of the village by killing a sheep and two bottles of schnapps. This charge is sufficiently deterrent to scare people from infringing or breaking the taboos.

In theory, the use of IK in biodiversity conservation finds here one of its strengths. Maintenance of rules based on tradition is stronger and more community owned than governmental rules. This offers an opportunity to involve people in biodiversity conservation using their indigenous knowledge systems. However, Ghana is a country in fast developmental transition. This applies to the length of modern roads, but also to the way traditional knowledge is used and applied. Education brought a lot of scepticism about traditional customs and beliefs. The situation described for agriculture where farmers find themselves in between the traditional and the Western approach, in a situation that can only be maintained at a high societal and environmental cost, also applies the most

Table 3: Instruments and Systems Underlying IK Actions that are Biodiversity Related

Instrument/System	Examples
Taboo and prohibition	Clearing vegetation along riverside.
	Days and periods when fishing or hunting is prohibited.
Respect traditional spirits	Sacred groves.
	Trees that should only be felt after accomplishment of rituals.
Common belief	Cultivate land in such a way that it can rest afterwards during a period that is long enough for the recuperation of the fertility of the soil.
Land tenure	Land owned as common, communal, clan or extended-family property.

other areas where interesting aspects of biodiversity related indigenous knowledge can be detected. Hunting taboos, even when they are totem related, are not maintained anymore. Bushmeat from hunted animals can be bought all the year around. In Wly it is easy to buy a bat hunted in the protected area near the waterfall for approximately 5000 cedis (0.50 Euro) per specimen.

Therefore, the possible effects of IK on biodiversity conservation are real, but should not be overestimated. Figure 1 shows a series of problems and their consequences as they are common today in many locally managed environments. Slash and burn practices coupled with fast rotation, hardly controlled wildlife hunting, destructive fishing using explosives or poisons, illegal logging of trees and ecosystems that are worth protection, fuelwood collection and charcoal production in areas that are less appropriate for these activities, over-collection of Non-Timber Forest Resources (NTFR) as snails, mushrooms, honey, fruits and land use to provide the increasing population shelter. All this is only possible at a tremendous environmental cost. This entails erosion, soil degradation both in quality and quantity, reduced agricultural, hunting and collection yields, degraded forests and depleted water resources.

When comparing the potentials on biodiversity conservation by IK, with the everyday life situation on the field, this points to at least a twofold conclusion:

- Attention should be given to document and inventarise traditional knowledge (related to biodiversity) as the available knowledge is at risk of disappearing fast.
- Education should give attention to the values (and limitations) of IK, so that it can obtain a fair place in the development paths of countries as Ghana.

The basics of astronomy, mathematics, pharmacology, food technology, metallurgy, zoology, botany and forestry at least in part, derive from traditional knowledge and practice. In contemporary science – although most of the listed disciplines have an ethno-component – learning from indigenous knowledge is the exception rather than the rule.

Figure 1: Problem Tree of Current Local Management Practices

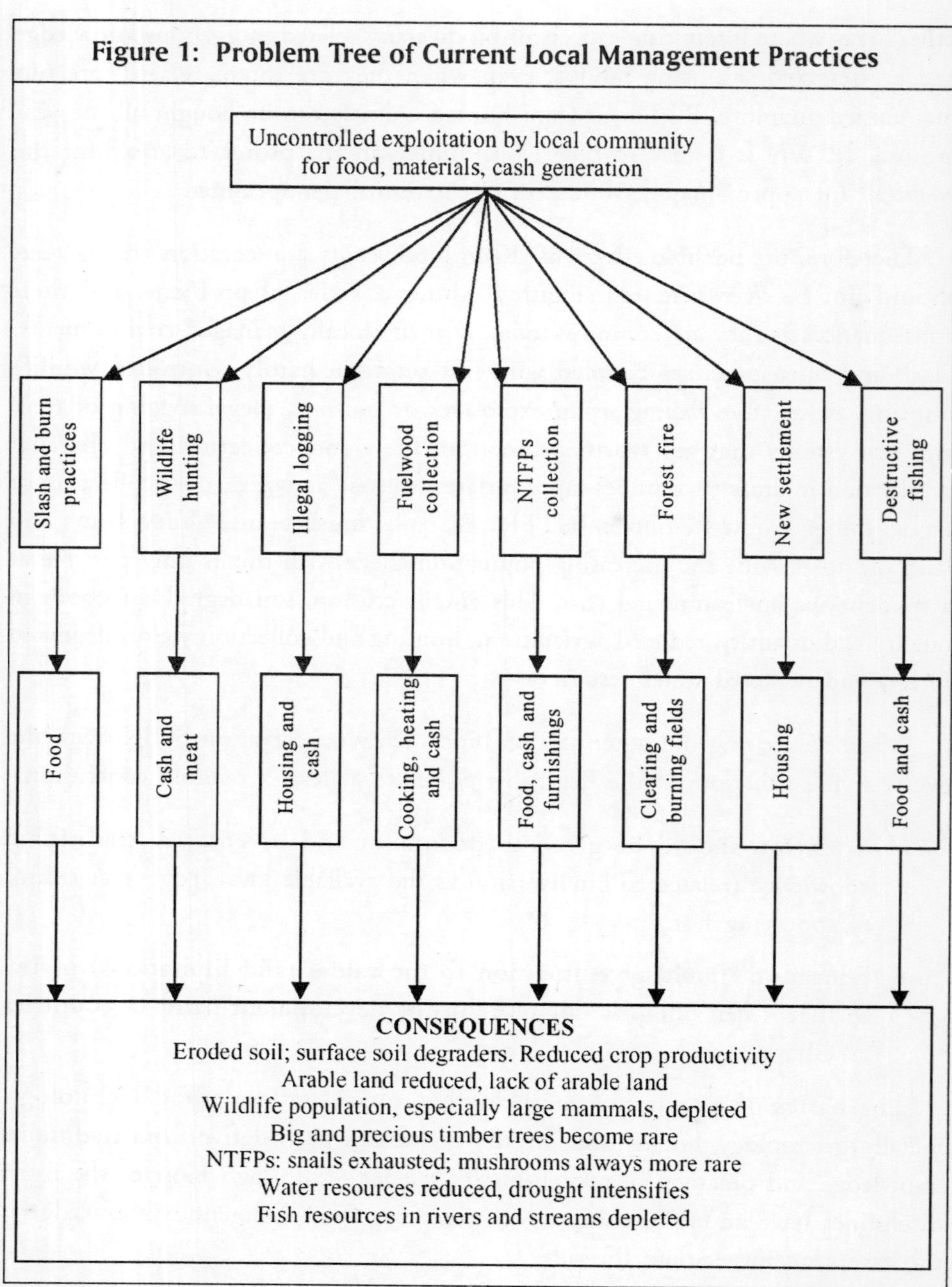

Table 4: Comparison of Selected Essential Characteristics of Indigenous Knowledge Systems and Contemporary Science

IKS	Science
Non-systematic approach	Systematic approach
Holistic approach	Reductionistic approach
Community based	Universal
Value loaded	Value free
Authority based	Counterdiction, verification, falsification, validation
Oral tradition	Written documents

There exist objective reasons for this finding. Table 4 compares essential characteristics of Indigenous Knowledge Systems (IKS) with their equivalents in science. The table shows how fundamentally different the paradigms in both knowledge systems are. While IKS cultivate an authority based, non-systematic, holistic approach; science is essentially systematic, driven by counterdiction, verification, falsification and validation and to a very large extent reductionistic. IKS have until now an oral tradition and are to a large extent community targeted; science on the other hand depends on written and published documents and strives to elucidating universal (both spatially and temporally) laws. Most sensitive in the tension between both systems is that indigenous knowledge roots in religion, occultism and transmitted beliefs. These essentials for IKS are alien to science.

3. Matching Indigenous Knowledge with Science and Technology

However, even at this paradigmatic level, both science and IKS are probably closer to each other than hardliners of both sides like to believe. IKS have their own systems that is often more detailed than the scientific one. Quaye (2005) collected in an ethno-botanical survey among 100 farmers in Upper West and Upper East regions of Ghana, data on the way they classify their plants. The local farmers growing Arachis hypogaea L., recognize varieties on the basis of morphological, ecological, agronomic and even culinary characteristics. The results of this study show that a more detailed and informative approach than the one installed by the Linnaean system is possible. Moreover this approach is important from a biodiversity point of view. The study gives credence to call for incorporation of indigenous knowledge in biodiversity conservation programmes in general and in crop germplasm conservation projects in particular.

Science, in particular its policy applications, has a strong local, regional character, just as IKS have. Moreover a holistic approach is not contradictory, to science. In particular in most biodiversity and human ecological issues, a holistic approach is even recommended. Science, as any human activity, might strive towards value free and universal statements and laws. The history of science shows it is not. Science, just as IKS, is characterized by the options in the subjects and questions that are researched (and those that are not). In medicine the attitude of the practitioner and the researcher is guided by the question: "What is the benefit for the patient?". Analogously, in biodiversity research the quality of the environment should prevail over industrial and other social agendas.

Also the sources of financing the scientists accept the socially and legally defined contract terms the researchers have to obey to, entail value components.

Nevertheless, essential differences exist. IK will only be more widely accepted and applied in policy when it can be matched with scientific data. That IK and modern science can complement each other, has until now, be most convincingly demonstrated in use of and research on herbal medicines.

Worldwide ethno-pharmacists have inventarised and mapped traditional, merely herbal medicines and their diagnoses. They have searched for new active substances and a number of them have been validated using the regular research phases a medical drug should pass. Often cited in this context is P57, a CSIR (Council for Scientific and Industrial Research – South Africa) patented pharmaceutical formulation for obesity control (Sibisi, 2004). The active substance is derived from a succulent of the genus hoodia. This plant was in rural South Africa used by the San for centuries as a substitute for food and water. The development of the drug took over 30 years. Main bottlenecks proved to be:

- The analytical chemical identification of the active substance. The research finally led to the discovery of a new family of molecules with anorexic properties.
- The discussion on licensing and royalties, which resulted in benefit sharing with the San.

Other examples are Gracino kola that has been shown to be effective in the treatment of glaucoma and hypertension. Extracts of the root of Xanthoxyllum zanthoxyloides improve the crisis in sickle cell anemia.

For biodiversity research the model to match science and indigenous knowledge is not as evolved as for the herbal medicines. Nevertheless, the essence of the research is simple: when a biodiversity resource is used, do not deplete it. Fish, but do not empty the river, the laguna or the sea. Hunt, but do not threat the species in their existence. Research should demonstrate to which extent IK rules are able to realize these targets of sustainable use. Complementary to this are the questions on effectiveness and efficiency of community based approaches to biodiversity conservation and management. In theory, when biodiversity can be maintained and monitored through customary laws, this is preferable over maintenance through codified law. However in fast modernizing societies where short-term profit often prevails over traditional customs, questions on effectiveness and efficiency should be at the core of the research.

4. Policy Implications

At an international level, the integration of IK in biodiversity policy has repeatedly being advocated. IUCN – The World Conservation Union – is a traditional leader of this international debate. It found its echoes in Rio's Agenda 21 and the Johannesburg Plan of Implementation. Also the Biodiversity Convention – that was opened for signature in 1992 and became operational one year later – initiated a debate and action on the issues related to the wider application of indigenous and traditional knowledge beyond local communities. All these calls are driven by the idea that IK is useful and valuable in natural resource management and that consequently, its wider application would be of benefit to the entire world.

However, as equity and justice are cornerstones in dealing with this issue, mechanisms should be found to guarantee a fair economic reimbursement for the wider application of IK to its owners. The clause in the Biodiversity Convention on royalties for the country that provided the species which was used to extract the recDNA of a new medical drug, is interesting in this respect, but is also felt as insufficient. As long as this issue is not solved in international agreements, there will be a resistance on the side of the developing countries to share their knowledge with industrialized countries, and they will continue to stress on mechanisms to keep this knowledge indoors.

But the problem is more complex. In 1998 the World Bank launched the Indigenous Knowledge for Development Programme. The activities focused on three key areas (a) raising awareness of the importance of IK; (b) enhancing local capacity to document and exchange IK; and (c) applying IK in development programmes. The evaluation (Gorjestani, 2004) mentioned four lessons learned and challenges. They are listed in Table 5.

Next to the intellectual property and previously discussed validation issues they stress on an interdisciplinary approach to IK. This refers to learning about the bio-physical environment, the social fabric, the local economy, culture and history, as well as the knowledge embedded in a community. Such a holistic approach can help to: better understand the local situation in its entirety; design more effective and sustainable programmes; learn from communities and help them to learn to adapt global practices locally. A main problem with this approach is that at least the scientific part has limited experience with integrating, holistic approaches. As mentioned before, contemporary science is in essence reductionistic.

Table 5: Barriers in Combining IKS with Policy (Gorjestani, 2004)
1. Learning about IK enables an interdisciplinary approach to development. 2. IK is highly context specific and as such not easily replicable unless leveraged with other knowledge systems. 3. Validation of IK is necessary to confirm replication. This requires protocols that fit the specific nature of IK. 4. New protocols to protect intellectual property associated with IK.

By definition IK is context specific. What works successfully in one location or for one community may not for another. The defeat is to extract from the knowledge that applies in a particular context, the more general aspects that can be applied elsewhere. Experience and case studies show that this is possible, but at the same time that it necessitates a careful approach.

While the discussion on the international policy front develops, governments in sub-Saharan Africa fail to incorporate IK in their policies. The "National Biodiversity Strategy for Ghana" (Ministry of Environment and Science, 2002) does not make any reference to IK. This supports the idea that during recent years substantial lip service has been given to the value of IK, but few – if any – concrete policy results emerged.

There is nevertheless an important and unmet need especially in Ghana to involve communities more explicitly in biodiversity conservation. Two examples might clarify this point: the conservation of the tropical forest and the current management of nature reserves. For decennia and since the colonial era, Ghana is an important exporter of tropical timber. However, today only small remnants of the tropical forest in the South Western part of the country have been left. The official policy of selective logging has failed in this respect that areas that were logged half a century ago cannot be logged again, as there are hardly commercially interesting trees. To avoid that the country will become an importer of timber within a relatively short period there is only one solution with two main components:

- Conserve the forest that is still left and halt the current selective logging policy of Ghana's tropical forest.
- Promote agro-forestry of indigenous species. To do so, the authorities urgently need partners among the villagers, chiefs and elders living in the logged areas. To realize such partnerships a proper land policy, microcredits and programmes raising additional income (e.g., through the commercialization of non timber forest resources) are needed. Next to these targeted instruments specific attention should be given to the general development condition of these deprived communities: their needs for sanitation, curative health services and social and educational amenities. This might be an important complement to the existing forest biodiversity policy of using environmental impact assessment, permits and other legal regulations, that has been proven to be unable to halt the degradation of what once was one of the countries main natural resources: Ghana's tropical forest.

  Kakun Conservation Area is a 360 km^2 large nature reserve near the Southern city of Cape Coast. The tropical park is precious because a population with an estimated number of 240 to 300 forest elephants is reported to live in the park. There are 51 communities living around the conservation area, which directly interfere with the biodiversity conservation policy in at least two ways:

- Farmers regularly report destruction of their fields near the conservation area by the elephants.
- Poaching of bush meat in the conservation area.

Currently, the communities are in no way involved in the management of the conservation area. They are just told to stay out of it. Nevertheless, there is an obvious need for partnerships with these communities that should go further and deeper than just refunding them for the reported damage to their fields. They should be educated on the importance of the reserve. The conservation area should gain more locally owned social acceptance. They should be involved in monitoring and maintenance of the park. They should contribute to the establishment of transition zones between the park and the communities. They should be involved in solutions to ban poaching. Today these community partnerships for biodiversity conservation is almost in existent.

5. Conclusion

This paper shows that IK can play a role in biodiversity conservation. Traditional water, soil and ecosystem management prohibitions on fishing and hunting and the ecological value of sanctuaries provide keys to involve traditional knowledge and practice in a more explicit way than before in biodiversity conservation and management. However, there are serious indications that with development and modernization, this knowledge and its application is lost fast. Therefore, it is urgent to collect and systematise this knowledge, taking into account the specific context in which it developed and exists. To avoid fossilization, IK should be part of the terms of reference of the school system.

Indigenous knowledge is not only at its best when it is matched with contemporary science; this match is also a necessary requirement to incorporate it in policy approaches. When it comes to IK on biodiversity this matching process probably is not that much in need of scientific validation (the forecasted effects are almost trivial), but raises merely questions on effectiveness and efficiency.

At the policy level, a most interesting international discussion has been initiated e.g. during the implementation of the biodiversity convention. A lot of this discussion culminates in the need to develop new approaches to protect intellectual property associated with IK. This is however insufficient. There is also a need to build capacity among local communities to develop, share and apply their traditional knowledge; to monitor the effects of actions based on IK, and for means to allow research in this area. There is equally a high unmet need to develop national strategies to support the use of IK. This should result in partnerships between local communities, governments, civil society and academia. It looks like today IK is at the beginning of a new cycle to learn from communities and helping communities to learn.

List of Abbreviations

CSIR	Council for Scientific and Industrial Research (of South Africa)
CSRPM	Centre for Scientific Research into Plant Medicine
IK	Indigenous Knowledge
IKS	Indigenous Knowledge System
TWCN	The World Conservation Union
NTFPs	Non-Timber Forest Products
NTFRs	Non-Timber Forest Resources
UNCED	United Nations Conference on Environment and Development
WSSD	World Summit on Sustainable Development

Acknowledgement

The author would most sincerely like to thank the following experts who provided the information for this paper:

- Dr. E C Quaye, University of Cape Coast.
- O Ampadu-Agepei, Conservation International Ghana.
- A Fallotey, EPA Accra.
- Agjei, Accra.p

(Luc Hens, Human Ecology Department, Vrije Universiteit Brussel, Laarbeeklaan Belgium. The author can be reached at human.ecology@vub.ac.be).

References

Alabi K A: *Traditional Forest Conservation in Practice: A Case Study of Ikire-Ile In Ola-Oluwa Local Government.* M. Sc. Degree Dissertation. Department of forest resources management. University of Ibadan, Nigeria (1992).

Amlalo D S, Atsiatorme L D and Fiati C: *Biodiversity Conservation: Traditional Knowledge and Modern Concepts. Proceedings of Tthe Third UNESCO MAB Regional Seminar on Biosphere Reserves for Biodiversity Conservation and Sustainable Development in Anglophone Africa (BRAAF).* Environmental Protection Agency, Accra, Ghana, p. 157 (1998).

Boaten I B A A: Traditional conservation practices -Ghana's example, pp. 1-6, In: *Biodiversity Conservation: Traditional Knowledge and Modern Concepts. D S* Amlalo, L D Atsiatorme, and C Fiati (Eds). Environmental Protection Agency, Accra, Ghana (1998).

Bonsu M, Apori S O, Quaye E C and Anderson T: *Indigenous Soil, Water and Forest Resources Conservation in Ghana: A Case Study of Brong-Ahafo and Western Regions.* Friends of the Earth, Accra, Ghana, p. 53 (2000).

Conservation International Ghana: *Handbook of Totems in Ghana,* Accra, Ghana, p. 36 (2005).

Gorjestani N: Indigenous knowledge: the way forward, pp. 45-55, In: *Indigenous Knowledge: Local Pathways to Global Development – Marketing Five Years of the World Bank Indigenous Knowledge for Development Program.* R Woytek, P Shroff-Mehta and P C Mohan (Eds.). Knowledge and Learning Group, Africa Region, and World Bank, IK Notes, Washington DC (2004).

Hens L, and Boon E K: *Causes of Biodiversity Loss: A Human Ecology Analysis.* VUB, Brussels (2000).

Hens L and Nath, B: *The Johannesburg Conference. Environment, Development and Sustainability* 5: 7-39 (2003).

Ministry of Environment and Science: *National Biodiversity Strategy for Ghana,* Accra, Ghana (2002).

Ola Adams B A: Traditional African knowledge and strategies for the conservation of biodiversity-prospects and constraints, pp. 33-42, In: *Biodiversity Conservation: Traditional Knowledge and Modern Concepts.* D S Alalo, L D Atsiatorme and C Fiati (Eds). Environmental Protection Agency, Accra, Ghana (1998).

Policy Research and Strategic Planning Institute: *Ghana Herbal Pharmacopoeia,* Advent Press, Osu, Accra (1992).

Quaye E C: Personal communication (2005).

Sam G H: Harmonizing traditional and modern knowledge and concepts of biodiversity conservation. The case of medicinal plants, pp. 76-78, In: *Biodiversity Conservation: Traditional Knowledge and Modern Concepts.* D S Alalo, L D Atsiatorme and C Fiati (Eds). Environmental Protection Agency, Accra, Ghana (1998).

Sibisi S: Indigenous knowledge and science and technology: conflict, counterdiction or convenience? pp. 34-38, In: *Biodiversity Conservation: Traditional Knowledge and Modern Concepts.* D S Alalo, L D Atsiatorme and C Fiati (Eds), Environmental Protection Agency, Accra, Ghana (2004).

Sinclair A R E: The loss of biodiversity: the sixth great extinction, pp. 9-15, In: *Conserving Nature's Diversity.* G C Van Kooten, E H Bulte, and A R E Sinclair (Eds). Ashgate, Vermont (2000a).

Sinclair A R E: Is conservation achieving its ends?, pp. 30-44, In: *Conserving Nature's Diversity.* G C Van Kooten, E H Bulte and A R E Sinclair (Eds). Ashgate, Vermont (2000b).

Steiner A and Oviedo G: Indigenous knowledge and natural resource management, pp. 30-33, In: *Biodiversity Conservation: Traditional Knowledge and Modern Concepts.* D S Alalo, L D Atsiatorme and C Fiati (Eds). Environmental Protection Agency, Accra, Ghana (2004).

Urena Hinojosa C J: *Diversidad Classification y Uso de Plantas Medicinales en la Communidad de Apillapampa de la Provincia Capinota del Departemento de Cochabamba.* Thesis Maestria in Ciencias Ambientales, Universidad Mayor de San Simon, Cochabamba, Bolivia, 106pp., (in Spanish) (2001).

Woytek R, Shroff-Mehta P and Mohan P C (eds.): *Indigenous Knowledge: Local Pathways to Global Development – Marketing Five Years of the World Bank Indigenous Knowledge for Development Program,* Knowledge and Learning Group, Africa Region, and World Bank, IK Notes. Washington DC, pp. 1-3 (2004).

11

Nature Makes them Lazy: Contested Perceptions of Place and Knowledge in the Lower Amazon Floodplain of Brazil

Mark Harris

This article considers how fisherpeople, who live on the Lower Amazonian floodplain, perceive their environment. It contrasts their views with those of elite town-people and attempts to put these differences in a brief historical context. If these perceptions are historical and part of an established tradition of local knowledge, what role should they play in the conservation of the floodplain? My aim in this article is to show how the floodplain is a multilayered place requiring skilled knowledge to survive. It is made by human labour, as well as the river and its movements. This kind of knowledge is a key resource for environmental management. By focusing on 'local knowledge' I am trying to complement the work of NGOs who are dedicated to community forms of management. My intention is to show the horizons that conservationists should be aware of if recent anthropological understandings of human-environmental relatedness are to be taken seriously. For all the labels and models used to

Source: Conservation and Society, Volume 3, No. 2, December 2005.

describe floodplain residents and their work we cannot really know them until we know what they know and how they come to know. This perspective complements the expert and specialised knowledge of outsiders.

1. Introduction

The Floodplain of the Amazon region (várzea) is a forested environment of the whitewater rivers—waters with a high level of sediment (Figure 1). The land is flooded seasonally due to rainfall and the melting snows of the Andes. The floodplain is so variable over its extent from Peru to the mouth of the great river that it is impossible to generalise and description must be limited to a micro region. The várzea comprises about 2% of the Brazilian Amazon and 12% of the Peruvian Amazon. At the mouth of the Amazon, the floodplain is affected by tidal movement, as well as seasonal fluctuations. In the Lower Amazon of Brazil[1]—the focus for this article—there is no significant tide, and the area is immensely flat, characterised by large shallow lakes, grasslands, forests, and a maze of channels and streams. What is fascinating about this environment is that people, animals and plants have adapted to the annual floods, which cover the entire landscape with water for several months each year; up to ten metres in some places. At high water or during the rainy season, their riverside houses, built on stilts, are islands against a vast flooded horizon; during the low water, the land is farmed. During the floods, cattle are kept on raised platforms and grass is brought to them. Fish move into the inundated forest and live on a diet of nuts and berries falling from the trees.

Until recently, the floodplain of the main Amazon river attracted little attention from scholars and research scientists. It covers a very small percentage of the area of the region, has little forest and appears to be a risky place to live given the uncertainty of the river's seasonal changes. What is more, it is inhabited mostly by mixed-blood poor peasants who held a marginal interest for anthropologists. In short, it seemed not to be threatened and relatively insignificant compared to larger scale schemes and concerns, such as road-building projects, land grabbing, Amerindian disease, forest loss, gold mining, etc. (Hecht and Cockburn, 1989). However, from the 1980s, this situation dramatically changed. There are now a number of scholarly studies and reviews and three important research-oriented

Figure 1: Map of the Amazon

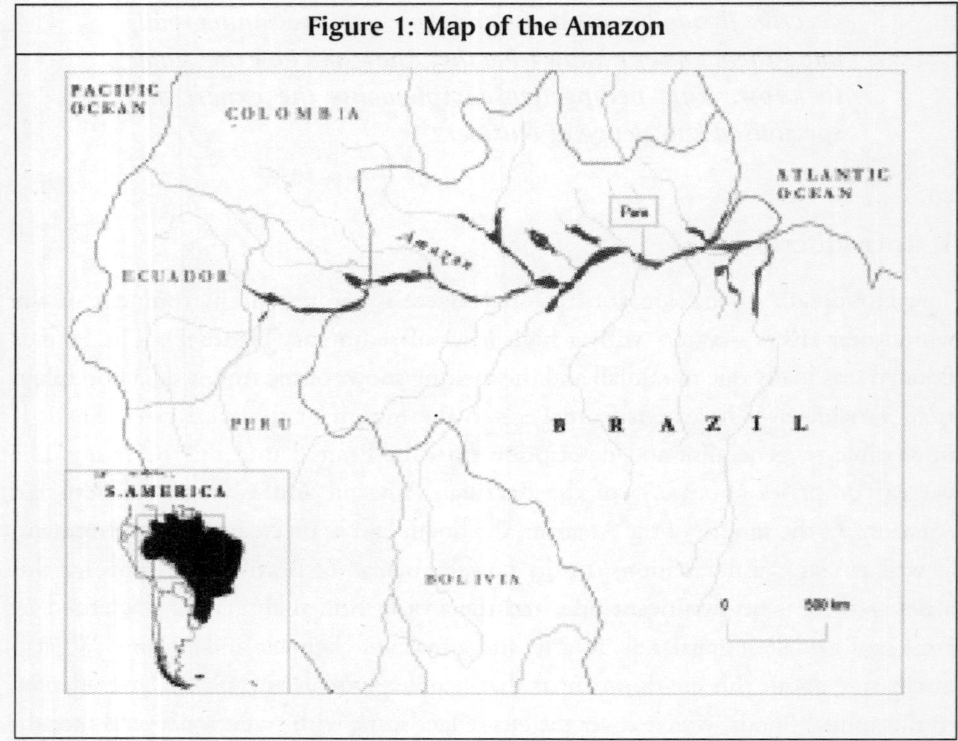

sustainable development projects dedicated exclusively to the floodplain environment (Mamirauá, IBAMA's Provarzea, and IPAM's Projecto Várzea, see websites in references). The reasons for this reversal in fortune are multiple. The research environment has opened up due to a paradigm shift, which has given the role of the floodplain in prehistory much greater importance. The floodplain has been identified as an important part of contemporary regional development. An increase in fish exports following the fall of other floodplain products has brought attention to the politics and science of its resources and management strategies (McGrath *et al.*, 1993). Various conflicts have resulted between locals and outsiders due to the intensification of fisheries and animal husbandry. The lack of legal recognition of floodplain dwellers' land, ownership has forced a reconsideration of how to conceive of land, which is flooded for half the year—whose property is it? In all, the floodplain of the Amazon has generated much scientific, governmental and civil society interest in the last twenty or so years. And at the centre are arguments about conservation and development.[2]

This article is concerned with the people who live on the floodplain, who are known as *ribeirinhos* in Portuguese, and are sometimes involved in these arguments. What role should their local knowledge of the floodplain play in these developments and conservation of the environment? The issue here is that this knowledge is not something that can be discovered through question and answer sessions and marked down with numbers and times and weights, though these measurements are an expression of it. It is rather a knowledge that is embedded in what people do and learn by repeated practice, such as how to throw a hand cast net. With long-term familiarity of the context and observation and imitation, understanding can be gained by a sympathetic outsider. Anthropologists are very well placed to make this contribution, since they spend long periods with the people they work with. This knowledge is taken for granted or the 'what goes without saying' of everyday life (Bourdieu, 1977) and which is nevertheless constituted by power relations. I have looked elsewhere at how these matters are internally stratified in terms of gender and age (Harris, 2005). My aim in this article is to show how the floodplain is a multilayered place requiring skilled knowledge to survive. It is made by human labour, as well as the river and its movements. This kind of knowledge is a key for resource management. My intention is to show the horizons that conservationists should be aware of, if recent anthropological understandings of human-environmental relatedness are to be taken seriously.

The need arises because of the urgency given to environmental dilemmas by governmental and Non-governmental Organisations (NGOs). In the case of the Amazon, there appears to be a perpetual crisis or state of emergency with regard to the future of the environment. Each crisis produces another which has a human impact. One does not doubt the significance of increased road paving, soya farming and illegal logging, to give three current examples. These activities are having huge effects on the region (e.g., Watts, 2005). However, the discourse of crisis and ongoing devastation legitimates expert knowledge rather than local knowledge, and external intervention rather than local empowerment. The perceived urgency given to environmental problems is associated with global discourses of nature and development. Surely it is for this reason that there has been a considerable growth in NGOs in the Amazon with finance from North America and Europe (Buclet, 2004). Most of the NGOs propose innovative models of development, stress local management of resources and the importance of conservation. However, Buclet (2004) has shown how they are nevertheless embedded in a social system that is imbued with traditional forms of domination and older paradigms of socio-ecological development. This context limits the full impact of their policies. While local knowledge does not have to be at odds with expert knowledge, it often is, given the local political realities in which the project takes place.

By focusing on 'local knowledge', I am trying to complement the work of NGOs who are dedicated to community forms of management, 'based on small holder practices, which optimise local resource production, while maintaining the integrity of local ecological processes' (McGrath *et al.*, 2005: 3). It is critical to place local knowledge in history (Harris, 2000; De Castro, 2002). Presentism, as a manifestation of urgency, is combated in a longer term perspective: the reconfiguration of the past to fit the concerns of the present. If we can see the here and now as but one present, rather than the present, then it is possible to have an enlarged understanding of how Amazonia's resources have been exploited by different classes for different markets over time. Amazonia has been a managed and human-made environment since well before conquest (Roosevelt, 1980; Raffles, 2002). Nugent (2000) argues that the historical realities of the region are undercut by global eco-politics agendas. The fact that nature is treated as a global resource means it is abstracted from concrete situations and local knowledge, effacing the inequalities between core and peripheral countries in the world system. Locating knowledge in ethnographically meaningful histories helps bring together the 'dwelt' (Ingold, 2000) in environmental and political structures.

History and Knowledge

A subsequent problem is the disparity between historically constituted practices and skills (i.e., embedded knowledge), such as fishing or cattle raising, and the context, which gives them a value or not. This value could be assigned by the market or by a development project. The importance of an activity or a kind of people can change with the influence of new possibilities and discourses. In an excellent critical review of recent developmentalist work in the Brazilian Amazon, Carneiro da Cunha and Almeida (2000) make a poignant argument in this regard. They see 'two major current misunderstandings' (Carneiro da Cunha and Almeida, 2000: 315) in the way developers have portrayed local people: (1) that traditional people are committed to conservation, as determined by Western concerns (2) that 'foreign' NGOs are responsible for the connection made between conservation of biological diversity and traditional people of the Amazon; making for an odd alliance between first-world activists and third-world leaders and left-wingers who share this belief (Carneiro da Cunha and Almeida, 2000: 315). The result of NGO involvement and outside discourse in local communities, they point out, is to make local folk neo-conservationist or neo-traditional (Carneiro da Cunha and Almeida, 2000: 335). 'Traditional populations' take on board the rhetoric of conservation and adapt their behaviour to fit in. There should be little surprise here. If a window of opportunity is opened, people are likely to climb through. This may not mean they change what they were doing before.

In the Brazilian Amazon there have been developmental programmes from the beginning of European conquest in the early 17th century (see De Castro 2002 for a complete account). First, missions had a charter to civilise Indians and make them into Christian vassals for the Portuguese Crown, but their worldwide organisations were dependent on local labour and production to financially support their orders. All this was controlled by royal charters and missionary regulations. A more explicit form of economic development was introduced in the 1750s after the expulsion of the financially powerful Jesuits. The Portuguese wanted tax revenue so they nationalised the means of commerce and instituted villages to be administered by a colonist and Indian chiefs. This failed, so by the early 19th century, the control of labour was the only means by which development could be achieved, therefore gangs of workers were set up. However, instead of working for the common good, they served the private interests of the elite. During the later part of the 19th century, the region was sold as a great paradise, healthy and with rich soils for farming and cattle-raising. But the rubber boom was dominant and it attracted poor labourers to tap the trees instead. The 20th century then saw these developments on a much larger scale with a series of half-hearted attempts to bring more settlers to the region and stimulate the economy, including the creation of one of the largest bauxite mines in the world in Porto Trombetas. Although this is a brief outline, it shows that projects of conservation take place in a well-trampled world of governmental, foreign and non-governmental initiatives. It is therefore not surprising that local folk have adapted their practices at various junctures in order to fit in with the prevailing orthodoxy.

Floodplain dwellers in the Lower Amazon have survived over this period because they have constantly adapted both to economic and ecological changes (Harris 2000, see also Lima and Alencar 2001). This does not mean they merely respond to external pressures and have no internal identity. Rather, they are able to hold on to who they are only by accommodating and adapting. I want to show that their capacity to do this has important implications for a dialogue between anthropology and conservation.

These comments have set up the significance of local knowledge in the Lower Amazon. This way of knowing is context-dependent but can be used in different contexts and has a range of applications. And yet, it is this knowledge of the world, which allows these people to reproduce, and its successful learning is critical. In other words, what counts as knowledge and for whom? And what kind of knowledge gives you what rights and can be represented by lawyers in a court of law or chamber of debating politicians? This article is neither about 'Traditional Ecological Knowledge' nor development; nevertheless some of these issues impinge on its subject matter. What right do the people of the floodplain, and others like them, have to be where they are? They do not have titles to their land and could, in theory, be expelled by someone who is able to force legal ownership through the courts. In the call for closer attention to skilled and embedded knowledge produced in certain historical realities and political relations, this article aims to show its relevance in mounting a defence of riverine dwellers' rights. A landscape belongs to those who know it.

Amazonia and Contested Natures

The following ethnographic material derives from my fieldwork in the Brazilian Amazon, which I have conducted since 1992 over a total of 33 months in five periods. The principal location of this fieldwork is a floodplain area known as Parú in the Lower Amazon where about 900 people live; it is about half a day's journey in a motorised boat from the town of Óbidos.

Conflicts over resources, most importantly land, have a long history in the Amazon (Hecht and Cockburn, 1989). Social resistance continues to have media attention, as seen in the landless peoples' movement (Movimento dos Sem-Terra) and land invasions (see Stedile, 2002). The focus of such conflicts has been the drive for modernisation and progress. Money, tax breaks, massive land grants, etc., have all been used to attract colonists to the region, often bringing them into direct confrontation with either indigenous people or peasant small holders whose system of land holding (based on usufruct rights) differs markedly from the bureaucratic state's recognition of ownership. On the floodplain of the Amazon, these conflicts have been less dramatic than elsewhere, but still significant. In recent times, there have been two kinds of disputes: over access to the productive lake fisheries and over land for cattle-raising (see McGrath *et al.*, 1993 for a review). McGrath and his colleagues' argument is that these conflicts on the floodplain have two possible outcomes. Either the floodplain dwellers will succeed in collectively controlling local resources (for example in the creation of an extractive reserve such as Mamiraruá, see Lima 1999a), or else they will fail and be pushed out by their inability to reproduce under such conditions and migrate to towns, leaving the land and lakes for large-scale cattle ranching and fishing.

The relevance of these conflicts to my discussion is to introduce the class based character of the contestation of resources and local knowledge. The phrase 'nature makes them lazy' should be understood within the context of different perceptions of the Amazon by those who wish to defend their control and access to resources, in this case, peasant fisherpeople, and those who wish to develop them in the name of progress, i.e., local elites. These differences have histories dating back to the first conquest of the region, and have been successively revived at key moments such as during the rounding up of Indians into mission villages (MacLachlan, 1973), or during the rubber boom (see Nery, 1981), and then during the military dictatorship (see Hecht and Cock-burn, 1989). What is significant is that these discourses all critically concern 'nature', understood as a diverse set of resources, what to do with it and who should do it. Not surprisingly, each class at different historical junctures has a different way of relating to nature. In other words, the developmentalist attention fails to appreciate the historical and cultural situatedness of labour. For this reason, responsible management strategies cannot be imposed simplistically upon complex historical realities (Nugent 1993, 2000).

The local history of Parú has been written about elsewhere (Harris, 2000)—essentially it is made up of various migrations of people, mostly detribalised and missionary Indians dating from the mid-eighteenth century, a few Northeastern

Brazilians who came to tap rubber from the mid-nineteenth century, and Italian farmers and traders who came at the beginning of the twentieth century. Within the twentieth century, there has been much regional movement as a result of marriage and search for land for growing jute. In other words, Parú is a heterogenous community, economically and socially differentiated.

I shall now present some representative anecdotes of my experiences of the middle class view of the floodplain dwellers in the region, which happen to come from the town of Óbidos, a town of colonial origin on the river in the Lower Amazon. The term has been traditionally used to refer to people from rural areas is caboclo, one which locally has pejorative overtones of backwardness, ineptitude and laziness (Lima, 1999b). The elite view derives from a longstanding discourse on nature in Amazonia, which essentially is imposed from the outside and concerns development and modernisation. In the rest of the article, I will develop the floodplain peasant view by way of contrast.

Óbidos is situated at the narrowest part of the Amazon river, about a mile across, and has a population of about 40,000, evenly divided in the 2001 census between urban and rural areas. Its main export products are Brazil nuts, cattle and fish. A hundred years ago it used to be among the most important towns on the river but in the twenty first century, its fortunes have deteriorated with the decline in river traffic. Few, if any, urban middle class people in the town of Óbidos understood why I wanted to live on the floodplain. Worse still, with those 'caboclos of the countryside'. They thought they knew all about the life of the caboclo on the floodplain. When they talked about caboclo life, they referred to the milk that flowed endlessly from cows, and to the fish, only waiting to be caught. In short, they spoke of the easy and languid life of the caboclo in the interior. For the élite in Óbidos, and here I deliberately turn to the present tense, the interior has a life totally different from their own, one that demands nothing; and thus the caboclo is lazy and unambitious, inferior and passive. They are poor, so the tune goes, and they are happy to be so, they know no better. This discourse is similar to the one used against indigenous Amazonians, because caboclos are the heirs of the Amer-indian environment (Ramos, 1991).

Soon after I arrived in Óbidos, I introduced myself to the people who ran a small museum, which held a collection of artefacts, some imported from Europe in the nineteenth century and some locally made. I talked with two ladies from rich Óbidense families. Having explained that I had returned from the floodplain that morning, they said how lovely it was during the dry season. I then explained my research and my intentions. This declaration was met with disdain. They commented that 'there is no need to help caboclos, because nature helps them. 'Look here,' the lady went on, 'the laziness of the caboclo of the region is a result of nature helping them too much and being too kind to them, the fish and birds are free. And because nature helps them, they don't need to work.'

Another incident with a lawyer further attests to the link between backwardness and the countryside. We talked in the small hotel in Óbidos. He asked me what I was doing. When I said I was living in the interior and doing fieldwork, he launched into an attack on the caboclo. 'The caboclos are lazy, with no desire for social mobility. All they do is catch fish enough to fill their stomach and then they are satisfied.' He went on to blame many of the region's problems, such as its lack of development compared to the rest of Brazil, on the caboclo's idleness. 'It is too easy to live here, and this lack of ambition is our biggest obstacle to development'. These are, in brief, middle class views and I present them also as subject positions just as real and embodied as the *ribeirinhos.*

Images linking abundance and indolence were echoed in many other conversations in the Lower Amazon. The perception that there is a profusion of natural resources means people do not have to work. For this reason, development cannot occur because there is no incentive for change, nothing to strive for. This idea of the 'Garden of Paradise' was also expressed by many of the nineteenth century commentators who travelled to the area, such as Henry Walter Bates and Alfred Russell Wallace (Cleary, 2001).

Not surprisingly, these views contrast with those of the people being referred to. In the following ethnography, a counterpart is created in opposition to the perceived laziness. The skills needed to survive and reproduce on the floodplain are carefully cultivated amongst groups of kinspeople. If such skills are not taught successfully then the means of life are put at stake. Before turning to this topic, let me deal with how the floodplain dwellers I know counteract the elites' representations of them as lazy and living off nature. This is a large and vague topic so I shall consider two elements: environmental legislation and ownership of resources. Though there is no resistance involved, there is a sense of holding firm against an unwanted force. A small example: the Brazilian environmental agency (IBAMA) has banned the sale of fish less than a certain size. The bosses who buy the fish in the processing factories have to enforce this regulation otherwise they would be fined (if found to have bought too young fish). Also certain predatory fishing methods (e.g., the Seine technique and more dramatically dynamite) are prohibited. The response of the floodplain dwellers is to say that these people know nothing of how fish live and reproduce or the other ways fishermen conserve fish stocks. 'How can they know when they do not fish themselves, all they have are their instruments and books' is how one man put it to me. It is not that this man was denying the overall decline in fish stocks but doubted the elite's attempts to check the fall. From the floodplain perspective, a superior vantage point is afforded from their 'working' knowledge. Indeed the general antipathy they feel for a gente fina, well to do people, relates to the latter's lack of production; they live off others, making money out of their goods and getting them to do the hard work. How dare they talk of laziness when they do not work the land and the water, peasants might well say. Work, for them is measured in terms of physical effort (see Harris, 2000).

This resonates with the peasant's tradition of ownership, especially land and other resources (see Harris, 2000; Lima, 2004). Work creates ownership, which is conceived in terms of rights to use. A person has usufruct rights to a plot of land if it is cleared, planted and harvested. If it is left fallow for more than five years it belongs to no one; in fact it returns to the ultimate 'owner', or creator (dono) of all life—God, according to floodplain dwellers. Ownership is not based in fixed rights and relations (secured through financial power), which is the model of the elites, but in continuous use and skilled activity. So a farmer does not own the land per se but its fruits. As long as the floodplain peasants remain on the land they maintain control of the natural resources. If they move off, they lose everything and the bosses and their large contingents of cattle and buffalo will move in. The topic of ownership of the floodplain and its complex legal conceptualisation is being tackled by Jose Benatti (2003).

Before continuing with the ethnographic exposition of local knowledge, it is necessary to give a framework for the perspective outlined here. The emphasis on local knowledge derives from work in ecological anthropology and developmental studies (Little, 1999; Pottier, *et al.*, 2003). In many ways, the term local knowledge is deceptive because it appears to be opposed to 'global knowledge'. I have argued elsewhere that the skills of the floodplain people are hybrids of local and global influences, mixtures of the new techniques from the outside and local innovation of them (Harris, 2005). Raffles (2002) has argued convincingly that the Amazon is made and unmade as a special place precisely because it has a complex configuration of outsider and insider discourses and practices over time. On a more general level, Bicker *et al.* (2004) have shown that indigenous knowledge of the environment is flexible, complex and variable. Thus, conservation and development projects, which attempt to incorporate community-based management systems are in a difficult position since it may be hard to pin down what people know without adequate understanding and training.[3]

'Re-placing nature' entails taking into account the ways in which humans generate and engage with their environments through historically specifiable skills. For humans learn how to dwell in environments through a process of maturation that involves the somatic work of the individual in a field of relations. As Merleau-Ponty puts it 'The world is not what I think, but what I live through' (1964: xvi–ii). This phenomenological perspective insists that living and knowing, and thus interpersonal relationships, are part of the same historical process (see also Toren, 1993). Knowledge, in this sense is relational (like the French connaître or German kennen; and can be distinguished from factual information, savoir or wissen). I want to suggest the difficulty in bringing together the 'dwelling perspective' (Ingold, 2000) and political realities fall away if a distinction between local and global history is not made (see Ingold's epilogue). Jakob Meloe has referred to skilled ways of knowing as a 'form of life' (drawing on Wittgenstein), 'forms of activities or practices, each with a history of its own' (n.d. 3). The form of life is not a foundation from which life proceeds but a flow, within which practices are situated. For Meloe at the heart of a form of life (n.d. 5), is a 'life of work', the physical effort of transforming resources for distribution and consumption.

An Exposition of Skilled Knowledge

Elsewhere (Harris, 1998), I have argued that seasonal variations on the floodplain should be seen as part of the creative movement of daily life. In this sense, annual environmental changes, such as the rising and falling of the river, fish migrations and plant growth and decay, do not determine social life. Instead, seasonality is constituted by the movements of people and the rhythmic structure of their activities, which resonate with and respond to periodic changes in the floodplain environment.

I shall develop this argument by looking at the perception and knowledge of seasonal changes, focusing on skill-based knowledge. Let me give a concrete example. Every few days, a member from each house that is stationed on the banks of the lower middle Amazon has to rebuild their port. This port consists of a number of planks of wood supported by cross bars and legs. It goes from land over the mud at the edge of the water, into the river itself. About five to seven metres long, it is an essential part of domestic life. Women wash clothes and pots on it, children clean fish on it, canoes are tied to it, everybody washes on it and incoming boats go up to it to drop people off and load them on. The port needs to be rebuilt every few days, or at most every week, in accordance with the rise and fall of the river. Nobody can have a port that is not properly functional, providing a safe and dry passage from river to land and back again. The men, who have to move the port by going waist deep in water and digging up the supports and then replacing them, have to be constantly aware of the changing river levels. And so does everyone else. There is then an ongoing perception and attention to changes in the river with the adjustment of the port and consequently people's lives. Even when people are not physically working on the port they remain connected to the river, registering its movement and any associated animal, fish and plant changes. The experience is embodied in practice, functioning as a permanent part of the self, creating an enduring connection between human and environment. It is possible, in this sense, to speak of perception and skills as 'pre-objective' (not pre-cultural), and emerging in a process of individual life development in a historically specific social and natural environment.

Building the port uses the senses to understand movement and change in the river. The sensual monitoring allows a person to know when and where to make the next move. This is a small, but significant example. The same connection between environment, activity and perception can be made for a range of other situations: reaping crops, where to fish, when to clear a field, and in hunting expeditions. And sometimes people get it wrong, the senses are not always reliable and the environmental changes not always predictable.

The ongoing perception of changes in the river level implies the active participation of the person in their monitoring. This kind of perception is an activity taking place between the world, the body and the head (Ingold, 2000). In this way it is the interaction between aspects of the environment, perceptual monitoring and cognitive interpretation of this information, which translates back into behaviour, creating a kind of continuous loop. As people move around, information is actively sought. The point is that skilled practitioners consult the world—what they perceive—simultaneously as they do their mental representations and beliefs.

The passing of time is experienced as the passing of definite activities. Each season, or part of a season, is known by the way people engage with environmental processes: what fish is migrating past the village, what food is available, what work needs to be done, who is available for work, who is working where, what techniques to use, what fish is selling well at the market and so on. People are stronger and healthier in the summer time. In winter, they get thinner and complain of being trapped in their houses as the floodwaters close in.

Similarly, the river is more than an object or a geographic aspect of this world. It expresses not just place but change. The river is a period of movement between one place and another and one time and another. It ploughs on downstream and rises and falls, it extends and endures. In the example of moving the house port, the river is not an object in the mind of the perceiving person but a subject in a continuous exchange between the body and the environment. The actions of rebuilding a house port are the result of the attention to river level changes, involving a whole bodily and collective re-adjustment of life. The river is a subject in the sense that a person sees with it, according to it, if you like (Ingold, 2000).

Simultaneously, the individual person is positioned in a web of relationships with other people, which has its own dynamic. Moving from seasonal engagement to the learning of skills, I follow Chaitlin and Lave's argument that 'there is no such thing as learning *sui generis,* but only changing participation in the culturally designed settings of everyday life' (Chaitlin and Lave, 1996: 6).

Indeed I discovered that this corresponds well to Amazon floodplain dwellers' understanding of how practical knowledge is reproduced. On one occasion I was watching some teenagers lassoing a young bull. They offered me the rope and invited me to try. After many attempts I was unable to get the loop around the neck of the bull. I pleaded with a man to give me some advice. But he was not forthcoming, saying that he was not a schoolteacher and I had to watch them some more and carry on practising. He was quite deliberate in his phrasing. Following this advice, I was then compelled to observe closely the whole complex of interactions between parents and children to understand the processes of education and communication. In the mornings the children were given orders to get water, clean this or that, fetch something from a neighbour, etc. These were the only times children were given instructions.

On the other hand, they were never taught formally how to fish, wash clothes, gut a fish or take the bones out of a fish, caulk a canoe, milk a cow, and so on. These activities were learnt by observation of elder kin and neighbours, and practice over and over again. In fact, when I asked whether these types of activities needed to be taught they were bemused by the idea. I understand now why children used to sit around, eyes huge and faces blank; that is when they were not playing with my pencil and paper. They might have appeared passive, but through the process of observing adult interactions and work, they were actively participating in their own development. Obviously, some things needed to be taught, such as asking for blessing from elders, as a mark of respect.

How to work in a team is another skill that is not taught on the floodplain of Parú. And yet on a daily basis men and women work together in activities such as fishing, rounding up cattle, etc. Co-operation is fundamental to the completion of the task and ultimately to survival and social reproduction. This means being constantly aware of what other people are doing, showing respect for individual autonomy and maintaining an ongoing adjustment of what one is doing in relation to the movements of others. Aside from the overall plan of a fishing expedition, rarely did I observe men discuss the division of labour or actual implementation of a task. They seemed to read into the actions and postures of each other and work in silent co-operation and communication. When I questioned men about how they thought such harmony was achieved, they replied that it was only possible after a period of working together. Sometimes you did fall out with a work partner if you did not get on well, in which case you would have to find another. This kind of communicative silence is a good illustration of what Csordas means by a somatic mode of attention (Csordas, 1993).

It is then not surprising that I could not be taught how to lasso or that my informants considered my way of paddling inferior to their own. Parúaros complain constantly that even though they would not substitute their floodplain way of life for the luxuries of urban dwelling, their lives are hard and harsh. They get all manner of diseases and suffer greatly in the flood time. Their way of life takes getting used to, *acostumado.* This habituation is borne of being there and having kin who live near you. It is an at-homeness, a being in the world, which is inter-subjective. It gives rise to an identity, which is located in the life-world rather its transcendence, as is the case when identity becomes abstracted to be associated with culture and politics. This form is eminently social, and it is part of the ordinary world, before it is fetishised, objectified or monopolised or controlled.

Acostumado also implies that it is possible to fall out of a field of skilled practitioners, such as when a person migrates to town and becomes involved in other relationships and activities. Thus the knowledge that people have of their environment, and the skills to exploit it, emerge from being in place; they are emplaced. This does not mean they cannot be transferred or abstracted to other places. They are learnt and acquired in place in a series of relationships. To be used to floodplain life means to enjoy the good life, as it is seen by people who live there.

For Parúaros, skills and practical knowledge are synonymous with control of one's personal life. Parúaros refer to this knowledge as *'nossa intelligência',* which loosely translates as 'our intelligence', but could also indicate a particular attitude. The phrase was used countless times in answer to my questions about why floodplain dwellers do this and that in such a way. There was no verbal explanation as to 'why cousins keep kin together', as they say, for example. Just a shrug of the shoulders and the statement *nossa intelligência'.* Following Bloch (1998), we can say this kind of pronouncement reveals the difference between implicit, non-linearly encoded knowledge and verbal, linear knowledge that is easily expressed in language. Intelligência refers to a certain understanding that is embodied in what people do and how they act. The ethnographer's perception of such local knowledge can only be arrived at through participation in the daily lives of the people concerned.

Conclusion

I have discussed the anthropological significance of the kind of knowledge that develops from living and being in a specific place and learning the skills necessary to survive. I have emphasised that much of this knowledge is just as much about the environment as social, political and economic life. I have tried to show how a historically contextualised and politically sensitive study is at the heart of this project. There is an autonomy in day to day decision-making, which has permitted a degree of freedom for rurally based floodplain peasant lives. In this space dwell the peasant skills and knowledge encountered in this paper.

As lacking in ambition or desire for progress as these floodplain people appear to developers, they are nevertheless carrying out a massive historical mission. Put simply, this task is the continued employment of skills, which allow for reproduction from one generation to another, representing hard won positions arising from specific social situations in the flow of historical activity. The apparent effortless employment of these skills should not be mistaken for laziness, as the Amazonian elites have done and still do. Rather it is the precise maintenance of order and reproduction in a world of chaos and change that is their achievement.

The overall point is illustrated by a recent comparative study of manioc cultivation by Stocker (2005) in the North and Northeast of Brazil. Manioc is a staple food of many people in Brazil; being economically significant, its work and products, are embedded in a cultural repertoire of food preferences, folklore and history. Stocker (2005) focuses on in situ maintenance of agrobiodi-versity through an analysis of local knowledge and management techniques. Following ethnobotanical criteria, she categorised 214 varieties of manioc, which differed according to their growing times, the quality of the flour produced, where the plant came from and aesthetic aspects of the plant. She juxtaposes this material with the knowledge of extension workers and research scientists who are either employed by the government or NGOs. The knowledge of these people is directed to encouraging modern practices and high-yielding varieties. Consequently they only recognise less than a handful of species of manioc and only deal with the technical aspects of growing crops. Stocker argues that the developers and extension workers are trained to increase production and the efficiency of marketing and distribution of a product, integrating the changes with ecological processes.

They are not trained to dignify local use of manioc by respecting local preferences, protecting the prices, minimise crop infestations or securing ownership of land. Without taking into account practical knowledge of these small scale farmers, agrobiodiversity and conservation is undermined. A change in thinking then is needed if small-scale farmers' and ribeirinhos' conservation practices are not reduced to a series of technical strategies for product improvement and modernisation. The baseline in the manioc farmers and floodplain communities is the intergenerational skills, which have helped them reproduce with little or no external intervention in the past few hundred years (De Castro, 2002).

Nature has not made *ribeirinhos,* and others like them, lazy, but has provided them with the conditions and resources to develop the skills and forms of livelihood to survive in successive epochs of an unequal world. Understanding those skills and how they are connected to social life and affected by history is the starting point for appreciating how they have managed life on the floodplain. *Ribeirinho* resilience is their great practical offering for conservation of the region.

Acknowledgements

This paper is based in fieldwork carried out from 1992 to the present, funded by the British Academy, the Economic and Social Research Council, the University St Andrews and the Leverhulme Trust. The ideas were first formulated during a postdoctoral fellowship awarded by the British Academy held at the University of Manchester. Apart from thanking these institutions for their support I also thank Ben Campbell for his editorial patience, astuteness and generosity, and the reviewers of this journal for their suggestions.

(Mark Harris, Department of Social Anthropology, University of St. Andrews, St. Andrews, UK. He can be reached at mh25@st-andrews.ac.uk).

Endnotes

1 This area is historically between the cities of Parintins, state of Amazonas, and Almeirim, state of Pará, on the main trunk of the Amazon river. The regional centre is Santarém and other smaller towns are Monte Alegre, Alen-quer, Óbidos.

2 The pioneering study is Hilgard O'Reilly Sternberg's analysis of a floodplain region near Manaus in the central Amazon first published in 1956 (1998). The geographer Smith's

(1981) study of fishing is the next significant social scientific study. Resource management analyses from a cultural ecological perspective are provided by McGrath *et al.* (1993) and de Castro (2002). Overviews include Hiraoka (1992), Goulding *et al.* (1995). Padoch *et al.* (1999). Junk (1997) has led the ecological understanding of the floodplain and especially the 'flood pulse' dynamic. Meggers (1996, original 1971) should also be credited with acknowledging the importance of the floodplain in preconquest societies. Roosevelt (1980) transformed the prehistorical view of the basin is Parmana by showing that the floodplain supported much larger more complex populations than was previously thought. Porro's (1995) ethnohistorical reconstructions have also provided evidence of large scale hierarchical societies at the time of conquest.

3 At the heart of local knowledge is skill, the capacity to enact learnt bodily actions. The theoretical reorientation centres on the nature of skill as being the marker of difference in human societies. Skill is here understood as a continually adjusting action involving brain, body, tool (if used), environment and social relation (Ingold 2000).

References

Benatti J H 2003, Direito de propriedade e proteção ambiental no Brasil: Appropriação e o uso de recursos naturais no ímovel rural. PhD dissertation. Universidade Federal do Pará/Núcleo de Altos Estudos Amazônicos. Belém

Bicker A, P Sillitoe and J Pottier (eds.) 2004, *Development and Local Knowledge: New Approaches to Issues in Natural Resources Management, Conservation and Agriculture.* Routledge, London.

Bloch M 1998, *How We Think They Think.* Westview, Boulder.

Bourdieu P 1977, *Outline of a Theory of Practice.* Cambridge University Press, Cambridge.

Buclet B 2004, Le Marché International de la Solidarité: Les Organisations Non Gouvernementales en Amazonie Brésilienne. Ph.D. thesis. EHESS. Paris.

Carneiro da Cunha M and M Almeida 2000, Indigenous people, traditional people and conservation in the Amazon in Brazil: Burden of the past, promise of the future. *Daedalus, Journal of the American Academy of Sciences* 129(2):315–338.

Chaitlin S and J Lave 1996, *Understanding Practice: Perspectives on Activity in Context.* Cambridge University Press, Cambridge.

Cleary D 2001, Science and the Representation of Nature in Amazonia: From La Condamine Through Da Cunha and Roosevelt. In: *Diversidade Biologica e Cultural da Amazonia.* (eds. M.A. D'Incão, I.C.G. Vieira, J. M. Cardoso da Silva and D. Oren), pp. 273–296 . Museu Paraense Emilio Goeldi/MCT, Belém.

Csordas T 1993, Somatic modes of attention. *Cultural Anthropology* 8(2):135–156.

De Castro F 2002, From myths to rules: the evolution of local management in the Amazonian floodplain. *Environment and History* 8:197–216.

Goulding, M N J H Smith and D J Mahar 1995, *Floods of Fortune: Ecology and Economy along the Amazon.* Columbia University Press, New York.

Harris M 1998, The rhythm of life: Seasonality and sociality in a riverine village. *Journal of the Royal Anthropological Institute* 4(1):65–82.

Harris M 2000, *Life on the Amazon: The Anthropology of a Brazilian Peasant Village.* Oxford University Press, Oxford.

Harris M 2005, Riding a wave: Embodied skills and colonial history on the floodplain of the Amazon. *Ethnos* 70(2):197–219.

Hecht S and A Cockburn 1989, The Fate of the Forest. Penguin, Harmond-sworth.

Hiraoka M 1992, Caboclo Resource Management: A Review. In: *Conservation in the Neotropics* (eds. K. Redford and C. Padoch), pp. 134–157. Columbia University Press, New York.

Ingold T 2000, *The Perception of the Environment: Essays on Livelihood, Dwelling and Skill.* Routledge, London.

IPAM Instituto de Pesquisa Ambiental da Amazônia, *http://www.ipam.org.br/* Junk W J 1997. *The Central Amazon Floodplain: Ecology of a Pulsing System.* Springer, New York.

Lima D 1999a, Equity, Sustainable Development and Biodiversity Conservation: Some Questions on the Ecological Partnership in the Amazon. In: *Vár-zea: Diversity, Development, and Conservation in Amazonia's Whitewater Floodplains,* (eds. C. Padoch, M. Ayres, M. Pinedo-Vasquez, and A Henderson), pp 247–263. The New York Botanical Garden Press, New York.

Lima D 1999b, A construção histórica do termo caboclo. *Novos Cadernos de NAEA* 2(2):5-32.

Lima D 2004, The Roça Legacy. In: Some Other Amazonians: *Modern Peasan-tries in History* (eds. S Nugent and M Harris), pp. 12–36. Institute for the Study of the Americas, London.

Lima D de M and E F Alencar 2001, A lembrança da História: memória social, ambiente e identidade na várzea do Médio Solimões. *Lusotopie* 2001:27–48.

Little P 1999, Environments and environmentalisms in anthropological research: Facing a new millennium. *Annual Review of Anthropology* 28:253–284.

MacLachlan C 1973, The Indian Labour Structure in the Portuguese Amazon (1700–1800). In: *The Colonial Roots of Modern Brazil* (ed. D Alden), pp. 199–230. University of California Press, Berkeley.

Mamirauá, (Instituto de Desenvolvimento Sustentável Mamirauá) *http://www.mamiraua.org.br/*

McGrath D, F Castro, C Futemma, B D Amaral and J Calabria 1993, Fisheries and the evolution of resource management on the Lower Amazon floodplain. *Human Ecology* 21(3):167–196.

McGrath D, O Almeida, M Crossa, A Cardoso, M Cunha 2005, Working towards community-based management of the Lower Amazon floodplain. *PLEC News and Views* New Series, Number 6, March.

Meggers B 1996, *Amazonia: Man and Culture in A Counterfeit Paradise* (Revised). Smithsonian Institution Press, New York.

Meloe J n d Remaking a form of life. Manuscript, University of Tromso. Merleau-Ponty, M. 1964. *Phenomenology of Perception.* Routledge, London.

Mintz S 1989, *Caribbean Transformations.* Columbia University Press (Morningside Edition), New York.

Nery B d S -A, 1981, original 1899. *The Land of the Amazons* (Translated from French by G Humphrey). Dutton, New York.

Nugent S 1993, *Amazonian Caboclo Society: An Essay on Invisibility and Peasant Economy.* Berg, Oxford.

Nugent S 2000, Good Risk, Bad Risk: Reflexive Modernisation and Amazo-nia. In: *Risk Revisited* (ed. P Caplan), pp. 226–248. Pluto, London.

Padoch C, M Ayres, M Pinedo-Vasquez and A Henderson (eds.) 1999, *Várzea: Diversity, Development, and Conservation in Amazonia's White-water Floodplains.* The New York Botanical Garden Press, New York.

Porro A 1995, *O Povo das Águas: Ensaios de Etno-História Amazônica.* R.J, Vozes, Petrópolis. Provarzea/IBAMA, *http://www.ibama.gov.br/provarzea/*

Pottier J, P Sillitoe, A Bicker (eds.) 2003, *Negotiating Local Knowledge: Identity and Power in Development.* Pluto, London.

Raffles H 2002, *In Amazonia: A Natural History.* Princeton University Press, Princeton.

Ramos A R 1991, Hall of mirrors: The rhetoric of indigenism in Brazil. *Critique of Anthropology* 11(2):155–177.

Roosevelt A 1980, *Parmana: Prehistoric Maize and Manioc Susbistence along the Amazon and the Orinocco.* Academic Press, London.

Smith N J H 1981, *Man, fishes and the Amazon River.* Columbia University Press, New York.

Stedile J P 2002, Landless battalions. *New Left Review* 15:77–104.

Sternberg, Hilgard O'Reilly. 1998. *A Agua e o Homem na Várzea do Careiro.* Two Volumes. Museu Paraense do Emilio Goeldi, Belém.

Stocker P 2005, Family Farmers and Manioc in Contemporary Brazil: The Management of Agrobiodiversity and Change. Ph.D Thesis. University of Edinburgh.

Toren C 1993. Making history: The significance of childhood for a comparative anthropology of mind. *Man* (n.s.) 28(3):461–478.

Watts J 2005, A hunger eating up the world. *The Guardian,* London. 10th November, 2005. *http://www.guardian.co.uk/china/story/0,7369, 1638858,00.html.*

Amazonian Biodiversity: Bio-Piracy Issues

The tropical rain forest in the Amazon region covers about 8 million square kilometers. The region also contains some of the most diverse life forms on earth. Its biodiversity has made a remarkable contribution to local, regional, and world economies. Biopiracy, which includes the smuggling of diverse forms of flora and fauna, the appropriation and monopolization of traditional population's knowledge and biological resources, causes the loss of control of traditional populations over their resources. In the Amazon region this is reportedly a very common phenomenon. This joint knowledge must not be seen as a commodity that can be sold or bought. In the last years, through the advance of biotechnology, the facilitating of registering international trademarks and patents as well as international agreements on intellectual property, such as TRIPs, the possibilities of such exploitation have multiplied. However, there are also efforts to revert this situation:

In 1992, during Eco-92 in Rio de Janeiro, the Convention on Biological Diversity was signed. The Convention aims, among others, at the regulation of access to biological resources and at sharing, in a fair and equitable way the results of research and development and the benefits arising from the commercial utilization of genetic resources with the communities.

In 1995, Senator Marina Silva (from 2003 on, Minister of Environment of Brazil) presented a law project to create legal mechanisms for practical arrangements according to the Convention on Biological Diversity.

In December 2001, shamans of various indigenous peoples from Brazil formulated the Letter from São Luis do Maranhão, an important document for the World Intellectual Property Organization, questioning all forms of patents that derive from traditional knowledge.

In 2002, ten years after Eco-92, the workshop "Growing Diversity" took place in Rio Branco, Acre. The event was organized by the international NGO GRAIN in partnership with the NGO GTA-Acre. 100 representatives of farmers, fishermen, indigenous people, extrativists, craftsmen and NGOs of 32 countries of Asia, Africa and Latin America, participated in this event. They formulated the "Commitment of Rio Branco", alerting the world on the serious threat of biopiracy and requiring, among others, that life patents and any form of intellectual property on biodiversity and traditional knowledge should be banned.

However, these efforts seem shy, compared to the greed of the speculators and multinational companies that continue to take over and monopolize the wealth of the Amazon.

Source: http://gefweb.org/Outreach/outreach-Publications/Project_factsheet/LAC-regi-3-bd-undp-eng.pdf http://www.amazonlink.org/biopiracy/index.htm

12

River Rehabilitation for Conservation of Fish Biodiversity in Monsoonal Asia

David Dudgeon

Freshwater biodiversity is under threat worldwide, but the intensity of threat in the Oriental biogeographic region of tropical Asia is exceptional. Asia is the most densely populated region on Earth. Many rivers in that region are grossly polluted, and significant portions of their drainage basins and floodplains have been deforested or otherwise degraded. Flow regulation has been practiced for centuries, and thousands of dams have been constructed, with the result that most of the rivers are now dammed, often at several points along their course. Irrigation, hydropower, and flood security are among the perceived benefits. Recent water engineering projects in Asia have been exceptionally aggressive; they include the world's largest and tallest dams in China and a water transfer scheme intended to link India's major rivers. Some of these projects, i.e., those on the Mekong, have international ramifications that have yet to be fully played out. Overexploitation has exacerbated the effects of habitat alterations on riverine biodiversity, particularly that of fishes. Some fishery stocks have collapsed, and many fish and other

vertebrate species are threatened with extinction. The pressure from growing impoverished human populations, increasingly concentrated in cities, has forced governments to focus on economic development rather than environmental protection and conservation. Although legislation has been introduced to control water pollution, which is a danger to human health, it is not explicitly intended to protect biodiversity. Where legislation has been enforced, it can be effective against point-source polluters, but it has not significantly reduced the huge quantities of organic pollution from agricultural and domestic sources that contaminate rivers such as the Ganges and Yangtze. River scientists in Asia appear to have had little influence on policy makers or the implementation of water development projects. Human demands from agriculture and industry dominate water allocation policies, and in-stream flow needs for ecosystems have yet to be widely addressed. Restoration of Asian rivers to their original state is impractical given the constraints prevailing in the region, but some degree of rehabilitation will be possible if relevant legislation and scientific information are promptly applied. Opportunities do exist: enforcement of environmental legislation in China has been strengthened, leading to the suspension of major dam projects. The 2003 introduction of an annual fishing moratorium along the Yangtze River, as well as breeding and restocking programs for endangered fishes in the Yangtze and Mekong, offer the chance to leverage other initiatives that enhance river health and preserve biodiversity, particularly that of fish species. Preliminary data indicate that degraded rivers still retain some biodiversity that can be the focus of rehabilitation efforts. To strengthen these efforts, it is important to identify which ecological features enhance biodiversity and which ones make rivers more vulnerable to human impacts.

1. Introduction

When we think of charismatic and endangered megafauna, tigers, rhinos, and pandas immediately come to mind. With a little more consideration we could add some aquatic species to this list, such as whales, sea turtles, and albatrosses. Freshwater animals are less likely to be categorized as endangered, and none of them seem to have sparked the public interest to the same extent as threatened marine and terrestrial species. For instance, river dolphins receive much less attention than their marine counterparts, despite the fact that three of the five taxa of "true" river dolphins, i.e., those that never enter the sea, are endangered or critically endangered (IUCN, 2004). The Yangtze dolphin (Lipotes vexillifer Miller) is one of the most threatened mammals on earth, numbering fewer than 100 individuals, and both the Ganges and Indus subspecies of Platanista gangetica (Roxburgh) are endangered. These river dolphins share another characteristic: they are endemic to the Orient, i.e., tropical or monsoonal Asia. Crocodiles, river turtles, specialist river birds, and large freshwater fishes in the region are also increasingly rare and globally threatened by overharvesting and habitat degradation (Wei *et al.*, 1997, Dudgeon 2000a,b, Baird *et al.*, 2001, Hogan *et al.*, 2001, 2004).

Although it could be argued that the status of large, conspicuous freshwater vertebrates is not necessarily a good indicator of the status of all the taxa that share their environment, there are nevertheless grounds for serious concern about freshwater biodiversity in monsoonal Asia. A recent series of reviews (Dudgeon 1999, 2000a,b,c,d, 2002a,b) has underscored the alarming condition of the region's rivers, which has been apparent for over a decade (Dudgeon, 1992). The major rivers of the Oriental Region, including the Indus, Ganges, and Yangtze (Chang Jiang), have experienced centuries of sustained human impact and are among the most degraded, densely settled, and human-modified river basins on earth. Their waters are grossly polluted, and dams and impoundments influence their natural discharge to such an extent that the lower Ganges and the Indus virtually cease to flow during the dry season (Postel and Richter, 2003). Pressure from large impoverished human populations has forced most Asian Governments to focus on economic growth rather than environmental protection; the government response has been to implement massive development projects. There is an urgent need to halt ongoing habitat degradation and to restore or rehabilitate damaged Asian rivers or to manage them in a way that will sustain biodiversity.

This article examines the threats to riverine diversity in monsoonal Asia and describes how some of them can be addressed given the constraints arising from human use of water resources. Attention focuses mainly on fishes and the impacts of dam construction, with examples mainly from China, but this is not to imply that other taxa, threat categories, or countries are of less importance. Examples from elsewhere, e.g., Southeast and East Asia, are used when they offer informative parallels or contrasts with China.

Threat Categories

The gravity of threats to freshwater biodiversity in Asia results from the combination of three factors:

1. Asia supports a significant part of the world's biodiversity: for instance, Indonesia alone is home to about 15% of the word's species, and it has more amphibians and dragonflies than any other country (Braatz *et al.*, 1992).
2. Regardless of whether water is extracted, diverted, contained, or contaminated by humans, that use compromises its value as a habitat for organisms. As a result, most freshwater taxa are affected by a combination of threat factors, and this vulnerability to multiple interacting and often synergistic impacts puts freshwater biodiversity uniquely at risk among the Earth's biota.
3. Asia is the most densely populated region on Earth, with more than 50% of the human population in ~15% of the land area, and many people live in poverty. Five Asian countries account for about half of the global annual growth in population, and, although populations are becoming increasingly concentrated in cities, many people still live in rural areas, e.g., more than 70% in India. As a result, much of the landscape can be described as "human-dominated" (Hannah *et al.*, 1994). Deforestation and logging rates are the highest on Earth, and forest fires are frequent (Laurance, 1999, Taylor *et al.*, 1999; Achard *et al.*, 2002).

Threats to the biodiversity of Asian rivers and their associated wetlands include flow modification, habitat degradation, pollution, increased salinity, and overexpolitation (e.g., Dudgeon 1999, 2000b, c,d). In China, for example, ~46 X 109 t of urban and industrial wastewater are discharged annually; slightly less than half consists of industrial effluent, and about half of that is discharged into the Yangtze (SEPA, 2004). Until recently, less than 20% of the industrial discharge and municipal wastewater in China was treated; as a result, many rivers failed to meet government standards for drinking water supplies, and some were unsuitable even for agricultural purposes (Wang, 1989; SEPA, 1998; Wu *et al.*, 1999). The water along ~80% of the 50,000 km of major rivers in China was too polluted to sustain fisheries, and fish were entirely eliminated from at least 5% of the total river length (Wang, 1989; FAO, 1995; Wu *et al.*, 1999).

Contamination of water and risks to human health have prompted Governments throughout Asia to introduce legislation to control pollution (Dudgeon, *et al.*, 2000). These regulations are intended to ensure an uncontaminated supply of water for humans rather than preserve species or ecosystem goods and services. Furthermore, existing legislation generally has a negligible impact on reducing the huge quantities of organic pollution arising from agriculture and domestic sources. Legislative enforcement is weak in many areas (Dudgeon, 1999; Dudgeon *et al.*, 2000), but it can be effective against point-source industrial polluters, although control of small-scale village enterprises is limited. According to Low (1993:534) a major constraint is "... adequate manpower and funding, as well as the will to act ... which many countries lack. The financial and economic gains from unfettered development are too attractive to be hampered by enforcement of anti-pollution legislation."

Deforestation of drainage basins causes sedimentation, degrades rivers, and can have unexpected consequences for freshwater biodiversity (Brewer *et al.*, 2001). In addition, the conversion of floodplains and riparian zones to agriculture has detrimental effects on plants and animals in riverine wetlands (e.g., Dudgeon 2000c). Translocation of native species and exotic introductions are an additional threat to indigenous biota, and their influence as drivers of freshwater biodiversity loss is projected to increase substantially (Sala *et al.*, 2000), in part because exotic species are more successful in habitats that have already been modified or degraded by humans (e.g., Bunn and Arthington 2002; Koehn, 2004).

Flow regulation, which includes dam-building for hydroelectricity and impoundment of rivers to control floods and provide irrigation water, has a history of more than 4000 years in Asia, and its many effects range from the alteration of natural flow regimes to the obstruction of fish breeding migrations (e.g., Dudgeon 1995, 2000a). The ecological consequences of human-induced changes in flow variability are well established elsewhere (e.g., Poff *et al.*, 1997; Nilsson and Berggren, 2000; Bunn and Arthington, 2002); aggressive attempts to regulate flow are continuing over much of Asia. The monsoon climate causes highly variable natural flow regimes, and the dense human settlement of river floodplains has spurred the development of water engineering schemes for flood protection during high-flow periods and for water storage during low-flow periods and dry seasons. In China alone, for instance, floods kill thousands of people annually, and the 1931 floods in the Yangtze caused ~3 X 106 deaths.

Globally, dams retain ~10,000 km^3 of water: the equivalent of five times the volume in all the world's rivers combined (Nilsson and Berggren, 2000). The extent of water engineering in Asia is evident from the fact that 55% of the world's largest dams (> 15 m tall) have been built in China and India alone (WCD, 2000). Giant structures such as the Three Gorges Dam are under construction, and interbasin water transfers are planned, including a longstanding proposal to transfer water from the Yangtze to the more arid north of China (Dudgeon, 1995). In 2002, the Indian Government initiated a feasibility study for a scheme that would link India's major rivers by 2016 and transfer water from the north to the south of the county through 12,500 km of canals (Prakash, 2003; Bandyopadhyay and Perveen, 2004). Even if linking does not go ahead, global climate change is likely to be a driver of even more aggressive flow modifications in monsoonal Asia. Most change scenarios for the region predict an increase in climatic extremes and a greater frequency of floods and droughts (see Dudgeon, 2000a), all of which will demand water engineering responses.

In the face of ongoing threats and environmental degradation, the preservation of biodiversity will require the restoration or rehabilitation of Asian rivers. Both terms refer to the reparation or mitigation of the damage caused by human disturbance. More specifically, restoration means returning an ecosystem to its original structural condition, with its functional processes intact, by removing the causes of degradation. In contrast, the goal of rehabilitation is a partial rather than a complete recovery of ecosystem structure or function within the context of its present-day human use (FISRWG, 1998). Rehabilitation is a more realistic option in Asia, because it allows for management intervention with some degree of human disturbance. Accordingly, efforts aimed at improving the condition of Asian rivers are referred to in this paper as "rehabilitation." Given the degraded state of many of these rivers, almost any management intervention would contribute to their rehabilitation. Only in the relatively few places in which humans have had less impact, does river management in Asia constitute restoration to the river's original condition. This situation is rare, because management interventions seldom take place befores degradation is severe and fisheries are at or near collapse.

Has Pollution Control Been Effective?

To reduce water pollution in one of Asia's major rivers, the Indian Government initiated the Ganga Action Plan in 1985. The objective of this centrally funded scheme was to treat the effluent from all the major towns along the Ganges and reduce pollution in the river by at least 75% (Natarajan, 1989; Krishnamurti *et al.*, 1991; Payne *et al.*, 2004). The Ganga Action Plan built upon the existing, but weakly enforced, 1974 Water Prevention and Control Act. A government audit of the Ganga Action Plan in 2000, reported limited success in meeting effluent targets (Narayana Murty, 2000). Development plans for sewage treatment facilities were submitted by only 73% of the cities along the Ganges, and only 54% of these were judged acceptable by the authorities. Not all of the cities reported how much effluent was being treated, and many continued to discharge raw sewage into the river. Test audits of installed capacity indicated poor performance, and there were long delays in constructing planned treatment facilities. After 15 years of implementation, the audit estimated that the Ganga Action Plan had achieved only 14% of the anticipated sewage treatment capacity (Narayana Murty, 2000). The environmental impact of this failure has been exacerbated by the removal of large quantities of irrigation water from the Ganges, which offset any gains from effluent reductions.

In China, river pollution remains a serious problem, but significant steps have been taken to address the issue. In 2002, the Environmental Quality Standard for Surface Water (Standard Code GB 3838-88) was revised (GB 3838-2002), and enforcement measures against polluters were strengthened so as to make more effective use of the 1984 Law for Prevention and Control of Water Pollution.

A national network of sites for monitoring water quality in Chinese rivers was also established, although the number of monitoring stations varied from year to year, e.g., 741 in 2002, 407 in 2003. The State Environmental Protection Administration (SEPA), with financial support from the Central Government, initiated a series of key water pollution prevention and control projects across the country under the auspices of a series of national five-year plans for the prevention and control of water pollution. Enforcement of effluent control standards for industry and additional legislation introduced in 2002 intended to reduce pollution from large livestock-rearing operations have had some results. The proportion of discharged industrial effluent treated according to accepted standards is now close to 80% in some major cities (SEPA, 2004). Trends in the data acquired annually from the national network of monitoring sites provide evidence that river health is improving. Each site is classified according to national water quality standards ranging from Grade I (excellent) to Grade V (poor) and Worse than Grade V. The grading system is based on faecal coliform counts and, among other things, loadings of biological oxygen demand and chemical oxygen demand as well as the contents of nitrogen and phosphates (Table 1). The proportion of Grade I sites has increased slightly, and the percentage of those in the lower categories, particularly the category Worse than Grade V, has fallen. However, because of continued high levels of organic pollution and less stringent enforcement of effluent standards in small towns, more than half of the sites are still ranked as Grade IV or lower. There are particular problems in river tributaries, in which water quality is generally poorer than it is in mainstreams, and concerns remain over the effect of the accumulation of pollutants behind the Three Gorges Dam (SEPA, 2004).

Dams as Obstacles to Effective River Management

In densely populated Asia, multiple users compete for water resources. Conflicts can arise during the dry season and at other times of the year in areas in which water supplies are limited. Particular problems occur where rivers, such as the Ganges, Indus, and Mekong, traverse international boundaries. Activities upstream, e.g., pollution, water extraction, dam building, have important downstream implications for the natural transfer of energy and materials and the dispersal of pollutants, but the reverse is not true for most system characteristics. However, there are exceptions (see Pringle, 2001). Activities upstream can degrade

Table 1: The Percentage of Sites in China's National Network for River Monitoring that Met Specified Water-quality Standards or Grades During 2001, 2002, and 2003. The Intended Use of Each is also Shown. Raw Data were Acquired from China's State Environmental Protection Administration.

Classification	2001	2002	2003	Intended water use
Grade I	1.5	2.7	3.4	Drinking, nature reserve
Grade II	18.0	13.8	21.4	Drinking, fisheries
Grade III	10.0	12.6	13.3	Fisheries, recreation
Grade IV	17.7	18.9	23.8	Industry, noncontact recreation
Grade V	8.8	11.1	8.4	Agriculture
Worse than Grade V	44.0	40.9	29.7	None

the habitat and amenity value of downstream sections, and the potential for conflict is high because ~40% of the global human population lives in the 263 river basins that are shared by more than one country (Poff *et al.*, 2003; Postel and Richter, 2003). Indeed, negotiations over water sharing had a major influence on relations between India and Bangladesh following the construction of the Farakka Dam on the Ganges (Payne, *et al.*, 2004). Presently, there is international acrimony in South Asia over the construction of the Baglihar Dam on the Chenab River, a major tributary of the Indus in India (Hassan, 2005). Pakistan claims that the construction of this dam violates the Indus Water Treaty and will reduce the volume of water flowing into its territory.

One transboundary scheme that is likely to have very serious consequences for biodiversity involves the Mekong River. It is of special importance because of the relatively undisturbed and unpolluted condition of the river relative to the Indus or the Ganges. There has been no comprehensive biodiversity inventory of the Mekong Basin, but a recent estimated that 1,700 species of freshwater fishes inhabit the basin (Sverdrup-Jensen, 2002); this provides an indication of its richness. Even by a more conservative estimate of ~1000 species (Rainboth, 1996), it ranks third in the world for freshwater fishes next to the Amazon and Zaire Rivers (Dudgeon, 2002b). A substantial portion of the Mekong River flows through China, where it is known as the Lancang Jiang; it then flows south into Laos, Thailand, Cambodia, and Vietnam. China has ambitious plans for a cascade of huge mainstream dams on the Lancang Jiang. The Manwan Dam and Dachaoshan Dam, both more than 100m high, have already been built; the 300m Xiaowan Dam will be completed in 2010 (Figure 1). The Chinese portion of the Mekong contributes about 50% of the sediment load and 20% of the discharge at its mouth of the river, but most of the flow in Lao PDR and Thailand (Roberts, 2001a; TERRA, 2002).

Because of the enormous size of some of the Chinese dams, downstream effects on flows and sediment loads could be substantial. One prediction is that by 2010 the dams will reduce wet-season discharge and increase dry-season flows by 50% (Chapman and He, 1996), although other estimates project larger changes (TERRA, 2002). The consequence will be to "even out" the peaks and troughs of

Figure 1: Dams Planned or Constructed on the Lancang Jiang, the Section of the Mekong River that Flows through China Dates of Completion and Dam Heights are Shown

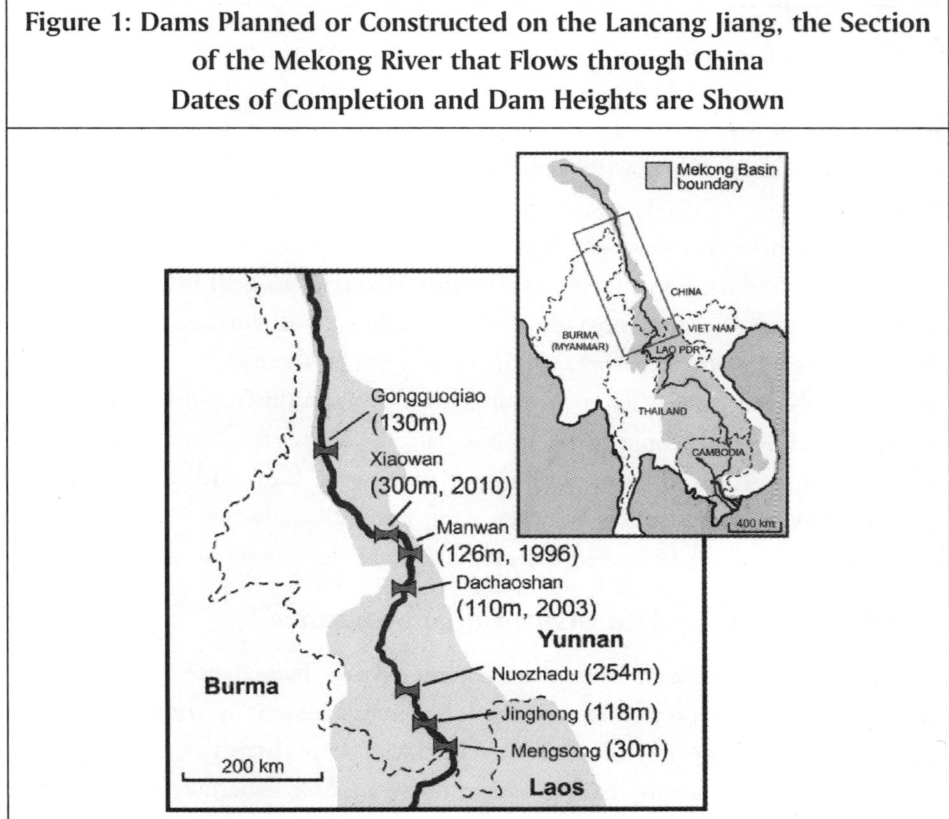

the natural discharge regime to which the Mekong River biota are adapted (for details see Dudgeon, 2000a). Unusually low dry-season flows in the lower Mekong during 2004, together with abnormal fluctuations in river level, have been attributed to the filling of the dams in China (Pearce, 2004). Changes in silt load because of sedimentation behind dams will have implications for riverbed erosion and agriculture downstream. Because these effects will be felt by the lower riparian states that have little to gain from the construction of dams in China, the scene is set for international conflicts of interest. Potential impacts on fish ecology will be especially important in land-locked Laos (Roberts, 2001a) where the freshwater fishing industry provides the main source of dietary protein for the human population. In a related development, China also has plans to dredge portions of the Mekong and to dynamite shoals and rock bars as to facilitate passage of large vessels (TERRA, 2002, Dudgeon, 2003a).

Conflicts of interest over river water also occur within national boundaries. A major consequence of hydropower dams is that their impacts on livelihoods and biodiversity are felt mainly by people in rural riparian communities. The collapse of artisanal fisheries after the 1994 completion of the Pak Mun Dam on the Mekong's largest tributary in Thailand is a conspicuous example (Roberts, 1993, 2001b). Dramatic declines in fish stocks occurred in 1998 as a result of dam construction on another Mekong tributary, the Theun River in Lao PDR, despite prior assessments that it would degrade the aquatic habitat downstream and obstruct fish breeding migrations (Usher, 1996). In contrast to these local impacts, most of the benefits from dams such as low electricity costs, industrial development, and flood protection for people living in low-lying areas, are felt some distance away, especially in towns and cities. The disparity between local and regional impacts and benefits creates conflicts between rural and urban dwellers that are usually settled in favor of the latter because they live closer to centers of political power.

Needs (1): Appropriate Institutional Structures

Conflicts over water use and river management have the best chance of being resolved with minimal environmental damage when management is coordinated at an appropriate ecological scale, but institutional structures for drainage basin management are generally lacking or ineffective in Asia, especially where adjacent countries share drainage basins. Disputes over water resources have been an ongoing feature of Indian relations with Pakistan and Bangladesh (Postel and Richter, 2003). Remarkably, the Mekong River Commission (MRC) offers an instructive example of a management structure for an international river basin in what has historically been one of the world's most fractious regions. The MRC is an intergovernmental body with ministerial-level representation that was established in 1995 by an agreement among the Governments of Cambodia, Lao PDR, Thailand, and Viet Nam. It was created by the Mekong River Committee itself, a modified version of an international organization established by the four riparian states in 1957, which was intended to coordinate water resource development in the lower basin (for more information, see Dudgeon, 1992, 2003a). The 1995 agreement mandated international cooperation "... in all fields of sustainable development, utilization, management and conservation of the water and related resources of the Mekong River Basin" (MRC, 2002:4). This statement is significant because it eschews a utilitarian approach to developing one or two major economic opportunities, such as hydropower or irrigation, in favor of adopting the broader perspective of holistic management. As a result, the MRC cancelled its plans to build a 12-dam cascade on the Mekong mainstream, which would have had major impacts on the ecology of the Mehong River (Dudgeon, 2000a). Future developments in the Mekong Basin are to involve joint planning in the context of an overall Basin Development Plan initiated in 2002. The Basin Development Plan is intended "... to identify, categorise and prioritise the projects and programmes to be implemented at the basin level ..." (MRC, 2002:8) with regard to matters such as irrigation, watershed management, fisheries, hydropower, navigation, recreation, water supply, and flood management. The National Basin Development Plan units in each riparian country should ensure that stakeholder interests are represented in the overall Basin Development Plan, and that ministerial representation by the MRC for each riparian state on the Mekong River facilitate the transition of plans into action.

Although the MRC provides a model of an institutional structure for the management of an international river basin, it has the shortcoming that China is not included and has not signed the 1995 agreement. China is an informal "dialogue partner" or observer of the MRC, but the chances of China becoming a full member are remote, because membership would interfere with the Chinese plans to build dams along the upper Mekong and enhance navigability further downstream. Furthermore, although there has been excellent cooperation between the five riparian states in the lower Mekong Basin, political commitment to holistic management will be put to the test within the next 5 years as decisions made in the context of MRC programs within the Basin Development Plan are implemented.

Needs (2): Environmental Flow Allocations

A major focus of the Mekong Basin Development Plan is to formulate a set of rules for the sharing of water sharing among the countries and various end-users to ensure the sustainability of fisheries and aquatic ecosystems (MRC, 2000). This is significant because, in most disagreements over multiple uses of water, whether they are international or local, the allocation of water resources to maintain biodiversity and ecosystem function receives scant attention (Poff *et al.*, 2003). Until recently, little attempt has been made to address the issue of environmental flows in China or India (Tharme, 2003). However, growing awareness of this matter in China, which has arisen in part from concerns that the Yellow River (Huang He) no longer flows to the sea for a significant part of the year, has stimulated research on ecological and environmental water requirements that address water allocation between humans and the environment (Cui, 2001; Ni *et al.*, 2002; Shi and Wang, 2002; Zheng *et al.*, 2002). This activity can be seen, in part, as a reflection of a growing global consensus about the need to adopt holistic environmental water allocations that sustain entire riverine ecosystems by providing water levels or discharges that mimic natural hydrologic variability (e.g., Poff *et al.*, 1997; Richter *et al.*, 1997; Arthington and Pusey, 2003; see also Liu *et al.*, 2005). Consideration of the need to allocate environmental flows downstream of hydropower projects will contribute to the preservation of fish stocks, other aquatic biodiversity, and ecosystem services, thereby avoiding situations that pit local impacts against distant regional benefits. There are more than 200 methods available for assessing environmental flows (Tharme, 2003), but their relative success has not been fully evaluated, and regionally relevant models for Asia have yet to be developed. Holistic environmental flow methodologies based on explicit links between changes in flow regime and the consequences for ecosystem parameters have been developed recently in Australia and southern Africa (e.g., Arthington and Pusey, 2003; Arthington *et al.*, 2003; King *et al.*, 2003; Tharme, 2003) and used to evaluate alternative water-allocation strategies in drainage basins for which data are scarce or detailed field investigations are not practical. Such methodologies will be appropriate in Asia where, initially at least, environmental flow allocations will have to be based on limited data and supplemented by best professional judgment and risk assessment. Assuming that this approach is adopted, an environmental flow allocation can be treated as a hypothesis-driven experiment in ecological restoration (Arthington and Pusey, 2003; Poff *et al.*, 2003). Rigorous monitoring and evaluation must follow the implementation of each environmental flow allocation, and the results used to refine the initial allocation strategy and inform subsequent environmental flow practices (Poff *et al.*, 2003). This process of refinement is essential given the range of scales over which environmental flows need to be allocated in Asia: from huge mainstream dams in China to smaller dams on Mekong tributaries, and from river reaches downstream of interbasin water-transfer schemes to large- or small-scale irrigation withdrawals.

The science of environmental flows is not likely to develop rapidly unless countries put in place legislative requirements and policy commitments to provide for environmental flows when considering water resource development. This would need to involve a mandatory wide-ranging environmental impact assessment followed by a more focused study on environmental allocations to maintain ecosystem functions that support biodiversity; studies of this type normally use hydrodynamic models (e.g., Arthington *et al.*, 2003; King *et al.*, 2003) to predict the consequences of the many different possible alterations to the natural flow regime. No such detailed analysis has been undertaken for major dams in Asia, and even the downstream ecological effect of the giant Three Gorges Scheme was treated superficially in the environmental impact assessment for the scheme (see Dudgeon, 1995). One hopeful sign is that, at the beginning of 2005, the MRC commenced a flow assessment program based on the evaluation of the beneficial environmental, social, and economic uses of the river, including specific consideration of the need to allocate environmental flows (MRC, 2005).

A significant opportunity to incorporate environmental flows into legislation in China is the law on environmental impact assessment promulgated in September 2003, which decrees that construction work cannot begin before the State Environmental Protection Administration (SEPA) has approved environmental impact assessment reports on the proposed project. For some time, this requirement has been flouted, most egregiously by the Three Gorges Company

(TGC), a state-owned corporation that has been building the world's largest hydropower project – the Three Gorges Dam. The TGC is also responsible for the US $5 X 109 Xiluodu Dam on the Jinsha River, the main upper tributary of the Yangtze (Figure 2), as well as three more dams planned for the same section of the river. Upon completion, the Xiluodu Dam will be the second largest hydropower plant in China. In December 2004, the State Council of China reaffirmed the law on environmental impact assessments and, in January 2005, the SEPA ordered work on the Xiluodu Dam to stop until relevant environmental impact assessments had been undertaken and approved. Ancillary projects associated with the Three Gorges Dam were also halted. Following some procrastination by the TGC, the SEPA reported in early February (Shenzhen Daily, 2005) that work on the Xiluodu Dam had ceased pending approval of the relevant reports, including studies of the impact of the dam on fishes in the upper Yangtze. The TGC was also fined for breaches of regulations. Although the matter has yet to play out, this is an important instance of the SEPA exerting control on the activity of a state-owned corporation, and it indicates a toughening stance toward environmental degradation in China.

The brake on dam building in China comes at a time when the SEPA has increased enforcement activities on factories that pollute, and this suggests a change in attitude toward the national imperative to develop now and clean up later. This change may reflect an increasing awareness of the consequences that two decades of untrammeled economic growth, e.g., almost 10% in 2004, have had on air and water pollution in China (Marquand, 2005). Further evidence of a new attitude was seen in April 2004, when Prime Minister Wen Jiabo ordered the suspension of planning for a cascade of 13 dams on the Nu Jiang, i.e., the portion of the Salween River that flows through China, pending a review of their impact on this UNESCO World Heritage Area. Preparatory work for the first of these dams at Liuku in Yunnan Province (Figure 2) is already well under way. The fate of the dam array now seems likely to depend on reports to be submitted to the SEPA (Yardley, 2005).

Protection of Fish Biodiversity in China

A greater emphasis on environmental protection and pollution control may, over the long term, contribute to the rehabilitation of rivers in China and elsewhere, but more focused action will be needed to protect endangered freshwater vertebrates such as fishes, river dolphins, etc. Even though China lacks a truly comprehensive law on nature conservation (Xu *et al.*, 1999), it did enact the China Wildlife Protection Law in 1989 to protect rare and endangered species. Although the law is yet to be completely enforced, China has established a legislative framework for biodiversity conservation and a number of action plans have been initiated (for details, see Xu *et al.*, 1999). A fishery law dating from 1986 and revised in 2000 proscribes fishing of rare and precious aquatic animals (Xu *et al.*, 1999, Fu *et al.*, 2003), and a National Red Data Book for threatened Chinese freshwater fishes has been produced (Yue and Chen, 1998). It lists 25 species of Yangtze fishes for the first time, in addition to both species of sturgeon that occur in the Yangtze, i.e., the Chinese Sturgeon (Acipenser sinensis Gray) and the Yangtze or Dabry's Sturgeon (Acipenser dabryanus Dumeril), and the Chinese Paddlefish (Psephurus gladius Martens). These three fishes are globally endangered (IUCN, 2004), and classified as a grade-1 protected species in China, where it is now illegal to catch them (Yue and Chen, 1998). Restrictions on fishing on the Yangtze are important because this river formerly contributed to approximately 70% of China's freshwater catch of ~5 X 106 t/yr. Yields fell to half this value between 1954 and 1970, and continued overfishing, flow regulation, and pollution caused catches to decline further to ~100,000 t/yr (FAO, 1995, Dudgeon, 2002a; Fu *et al.*, 2003; Chen *et al.*, 2004).

Within the last decade, an interprovincial authority, the Administrative Commission of the Yangtze River Fisheries Resources, was established under the Fisheries Bureau of the Ministry of Agriculture to control illegal fishing activities involving the use of electricity, explosives, and poisons, and to protect fishery stocks in the Yangtze. In February, 2003 an annual fishing moratorium of three to six months was introduced to protect 8,100 km of the river, including 4090 km of the 6300-km mainstream and more than 4000 km of its tributaries (XNA, 2004). This measure builds on the well established and widespread practice of stocking Chinese rivers and lateral lakes with cultured fry of major carp species to maintain or enhance fishery yields (Fu *et al.*, 2003): for example, ~100 X 106 fingerlings were released in the Yangtze in 2004 (XNA 2004). Such stocking cannot be regarded as a measure that contributes greatly to biodiversity conservation, because it has implications for the genetic variability of indigenous

Figure 2: The Location of Existing and Planned Major Dams on the Yangtze Mainstream, Including the Three Gorges, Xiluodu, and Gezhouba Dams. The Site of the First Dam Planned for the Nu Jiang (Salweeen River) is also Shown

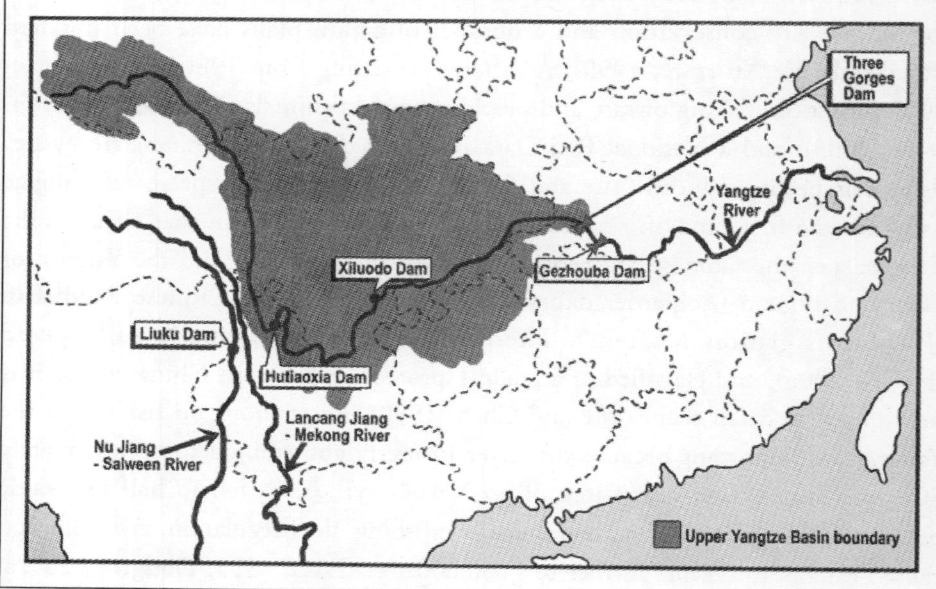

carp species. However, it may reduce harvesting pressure on rare species that could benefit from the fishing moratorium.

Stocking is not confined to major carp species. Since 1983, ~250,000 artificially propagated larvae of Chinese Sturgeon have been released into the Yangtze each year (Wei *et al.*, 1997, 2004). This stock enhancement began following the construction of the Gezhouba Dam on the Yangtze in 1981 (Figure 2), which blocked migrations, fragmented populations, and degraded spawning sites of sturgeons and Chinese Paddlefish (Dudgeon, 1995, 2000a; Wei *et al.*, 1997; Fu *et al.*, 2003; Chen, *et al.*, 2004). Concerns about the consequences of stocking on the variability of wild populations appeared to be justified when a genetic bottleneck was detected in samples of Chinese Sturgeon from the river (Zhang *et al.*, 2003). However, this might have been a result of a decline in their numbers, and loss of variability that occurred prior to 1983, which could have been a result of dam building, rather than an effect of inbreeding. Artificial propagation of Chinese Paddlefish has also been achieved with limited success. Reports indicate that 100,000 fry were released into the river in 2004 (XNA, 2004; but see Wei *et al.*, 2004). Stocks of the endemic and critically endangered Yangtze Sturgeon have not yet been introduced (Wei *et al.*, 2004). Genetic studies of wild Yangtze Sturgeon have shown a precipitous decline in diversity from 1958 to 1999, in parallel with a dramatic decrease in population size (Wan *et al.*, 2003); urgent management of the population is needed to preserve what variability remains. Notwithstanding concerns about the effects that artificially propagated individuals might have on viability or variability of wild populations of these rare fishes, the effectiveness of stocking as a conservation measure will be limited as long as the multiple threats of pollution, flow regulation, and habitat degradation persist in the Yangtze. Establishment of captive breeding populations of Acipenser spp. and Chinese Paddlefish as a sort of insurance policy against the loss of these fishes in the wild has been advocated (Yue and Chen, 1998; Wei *et al.*, 2004).

In the Pearl River (Zhujiang), China's second largest river, fishery stocks have suffered impacts and declines similar to those in the Yangtze (Liao *et al.*, 1989). The more than 3000 dams built since 1950 have blocked fish breeding migrations, leading to the virtual extinction of anadromous Reeve's Shad (Tenualosa reevesii Regan) in the river (Blaber *et al.*, 2003). This species formerly supported a highly lucrative fishery in both the Yangtze and the Pearl Rivers (Wang, 2003; Chen *et al.*, 2004). Although protective measures do not appear to be widely enforced, Reeve's Shad has been classified as a protected species by the Ministry of Agriculture since 1987. Successful pond culture of T reevesii has been initiated and may lead to stocking (Wang, 2003). All migratory Tenualsosa spp. in Asia are highly susceptible to human impacts resulting from overfishing, dam construction, and pollution. It is suspected that the endemic Mekong Shad (T thibaudaeui Durand) is near extinction (Blaber *et al.*, 2003), and the collapse of the once important Hilsa Shad (T ilisha Hamilton) fishery of the Ganges, following the 1975 completion of the Farakka Barrage, has been well documented (e.g., Natarajan, 1989; Chandra *et al.*, 1990); that collapse was predicted well before dam construction took place (Hora, 1942). According to Blaber *et al.*, (2003), declines have also been reported for both of the other species of Asian shad, the Longtail Shad (T macrura Bleeker) and the Toli shad (T toli Valenciennes).

Problems with protection of rare fishes and implementation of wildlife protection legislation in China exist because the natural resources laws and regulations, especially for fisheries, have been formulated from the standpoint of economic value, which emphasizes utilization rather than protection. Thus the conservation of endangered aquatic species is mainly the responsibility of the Ministry of Agriculture, which has had the historical remit of increasing or expanding the capture quotas of economically important species. Moreover, a separate authority, the Ministry of Water Resources, has an overlapping responsibility with the Ministry of Agriculture, but it is directed toward maintaining water supplies for human consumption and agriculture, whereas protection of water quality is under the jurisdiction of the State Environmental Protection Administration. The conflict between economic development and conservation plus the overlapping and sometimes contradictory responsibilities of government authorities have impeded actions that are needed to preserve biodiversity. Insufficient funding, a lack of trained manpower, and limited data sharing within China do nothing to improve matters (Xu *et al.*, 1999, 2000). Some progress has been made recently: in 2000, a 400-km section of the upper reaches of the Yangtze was designated as a reserve for rare fishes, but portions of it will be affected by rising water levels behind the Three Gorges Dam and by the dams planned for the Jinsha River tributary. Potential reserve sites for rare fishes elsewhere along the Yangtze have been identified (Fu *et al.*, 2003; Park *et al.*, 2003), but they await official designation.

Protection of Biodiversity in the Mekong

There are few laws protecting freshwater fishes in Asia, and reliable statistics for capture fisheries are lacking (FAO, 2000; Dudgeon, 2002a). However, there are a few exceptions to this generalization, as shown by some examples from the Mekong Basin. Cambodian law forbids the capture, sale, and transport of two endangered fishes that are endemic to the Mekong River: the Mekong Giant Catfish (Pangasianodon gigas Chevey), which is the world's largest freshwater fish (Figure 3), and the Giant Carp (Catlocarpio siamensis Boulenger). Nevertheless, enforcement remains problematic, and both species are still sold illegally at markets or to fish processing factories (Hogan *et al.*, 2001). Another Mekong endemic, the Smallscale Croaker (Boesemania microlepis Bleeker), is protected under a 1991 Lao PDR Ministry of Agriculture and Forestry decree that made it illegal to catch them during the spawning season or to sell them at any time of the year;

despite that, the sale and export of this species continue (Baird *et al.*, 2001). Severe fishing down of B. microlepis stocks to 10 to 20% of previous levels has stimulated the establishment of fish control zones by fishers in southern Lao PDR to protect deep-water spawning habitat; there is evidence of some stock recovery as a result (Baird *et al.*, 2001). It is worth noting that Sverdrup-Jensen (2002) considers fish control zones established by local communities to be more effective in protecting depleted stocks of large fishes than are legal restrictions on fishing methods and gear type, which are largely unenforceable. However, not all community-based fishery management schemes are conservation successes, in part because they may be directed toward maximizing fish yield (van Zalinge *et al.*, 2004). Local fish control zones have been established around deep pools in the Mekong in Lao PDR that are important because they serve as dry-season refuges for endangered species such as the Mekong Giant Catfish (Poulsen *et al.*, 2002). Protection of pools and spawning grounds could also favor large cyprinids threatened by overfishing. They include the genus Probarbus, which is represented by three species in the Mekong; two are endemic (Rainboth, 1996). Probarbus jullieni Sauvage (Seven-line Barb) has been eliminated from much of its former range in southeast Asia, e.g., by dam construction in the Perak River, Malaysia, and it is now globally endangered (IUCN, 2004); Probarbus labeamajor Roberts (Thick-lipped Bard) and P labeaminor Roberts (Thin-lipped Barb) are classified by the IUCN as data deficient, with the latter being the scarcer of these two Mekong species.

The Mekong Giant Catfish migrates from China to Cambodia and Thailand, where it is not protected. It once formed the basis of a small fishery in northern Thailand, but that fishery has now completely collapsed, and no Mekong Giant Catfish have been caught there since 2000 (Hogan *et al.*, 2004). Attempts to conserve the Mekong Giant Catfish in Thailand have included the release of artificially propagated fingerlings. Approximately 100,000 individuals have been stocked in the Mekong since 1985, but these fingerlings were obtained by sacrificing wild-caught fish to obtain eggs and sperm. Because very few adults are caught, there is a high degree of genetic similarity among their offspring, and the release of this stock might not only deplete the wild population in Thailand through the capture of brood stock, but also erode the genetic diversity of the Mekong Giant Catfish downstream in Cambodia (Hogan *et al.*, 2004). Furthermore, the perceived success of maintaining populations using artificial propagation of this threatened fish has delayed its listing as a protected species in Thailand (Nandeesha, 1994). Effective conservation of all migratory fishes in the Mekong is beyond the scope of the national legislative frameworks and will require coordinated action among the countries within the basin. At present, the Mekong River Commission offers the best framework for such cooperation (van Zalinge *et al.*, 2004).

What Role can Scientists Play?

Given the various human constraints, i.e., organization, legislative, and economic, what role can scientists play in the conservation of freshwater biodiversity in Asia? The provision of reliable information is an obvious contribution, especially if data can be collected over a sufficient period to indicate population trends. Unfortunately, fisheries data for Asian rivers are scant or of rather poor quality. For instance, although a considerable amount of research has been undertaken on Yangtze fisheries since the 1950s, most of it was directed by administrators, not scientists, with no comprehensive planning or consistent effort, and the resulting data are superficial, incomplete, and uneven (Chen *et al.*, 2004). In the Mekong, where the Mekong River Commission has taken responsibility for monitoring fisheries, there is no system for the effective collection of statistical data on landings. Existing official statistics grossly under-report catches, data on large-scale capture fisheries are inaccurate, and the authorities do not collect data on small-scale family fishing, because these fisheries have always been considered of minor importance to the national economy (Sverdrup-Jensen, 2002). Furthermore, there are very few quantitative data available for species and for habitats. The dispersed and varied nature of the fisheries in the Mekong makes the collection of accurate statistics problematic. Similar problems exist for other major inland fisheries in China, India, and Bangladesh (Sverdrup-Jensen, 2002). Along the Ganges, for instance, the fisheries are so spread out that " ... keeping representative statistics is always a difficult task" (Payne *et al.*, 2004:240).

Despite the lack of reliable data, the freshwater riverine and wetland fishery yields of the Mekong River are conservatively estimated at ~1533 X 106 t/yr worth US $1478 X 106, excluding yield reservoirs and aquaculture (Sverdrup-Jensen, 2002; van Zalinge *et al.*, 2004). This might make the Mekong the largest river fishery in the world, contributing to almost 2% of the global total for all

Figure 3: The Mekong Giant Catfish (Pangasianodon gigas) has Iconic Status in the Countries of the Lower Mekong Basin, Yet is Now Critically Endangered, Primarily Because of Overfishing

capture fisheries, i.e., ~ 90 X 106 t/yr (FAO, 2000). Given the magnitude of the Mekong fishery, the lack of reliable statistics is of great concern, because it limits both the management potential and our understanding of the factors that sustain fish productivity. Quantitative data on fisheries yields by species and by habitat are also required. Reliable information on the contribution of the fisheries sector to food security and income creation is necessary for a variety of stakeholders, including politicians and policy makers. For example, water resource planners need data that will allow them to evaluate the possible effects of development schemes on fisheries and assess the overall cost benefits of their activities. Fisheries data must also be collected and made available, and intelligible, to the general public because, unless the public understands the value of fisheries resources, management systems will not be viable (Sverdrup-Jensen, 2004).

Putting this paucity of reliable data on river fisheries into a wider context of information on freshwater ecosystems, one recent analysis shows that, between 1992 and 2001, scientists in Asia made a relatively minor contribution to the international literature on conservation and freshwater biology (Dudgeon, 2003a). Most of these papers concerned taxonomy, and many dealt with man-made lakes; here were scarcely any process-orientated investigations that could inform management. This doesn't mean that local and national contributions are unimportant or that the work was not worth doing. However, unless research results receive some exposure in international publications, they can have little impact on raising awareness of the need for conservation at governmental and global levels. Furthermore, if information generated by scientists in one country is not available internationally, it cannot be applied in other countries (Dudgeon, 2000d, 2003a). In particular, case studies that will help develop strategies for the restoration and rehabilitation of rivers and the establishment and assessment water allocations to maintain ecosystems are needed.

One point must be emphasized: we know that running waters in Asia have experienced sustained and severe human impacts and, in certain situations, can recover to the extent that they can support diverse aquatic communities (e.g., Dudgeon, 2003b). However, simply removing a threat, e.g., by controlling effluents, is no guarantee of restoration to the original pre-impact condition. A community quite different from those that developed in comparable reference systems, which were not subjected to the same disturbance may already have become established. Early colonization by exotics or predators may make it impossible for the original species to recover, or recovery may be impeded by some factor that affects the entire system, such as climate change or large-scale modifications of the characteristics of the drainage basin (Ormerod, 2004). In addition to studies of water allocation strategies, research is needed in at least three critical areas. First, the environmental factors that enhance species persistence in degraded Asian rivers must be identified. Second, the attributes of species that facilitate persistence or, conversely, increase the probability of extinction, must be determined. Third, management approaches that enhance the restoration or rehabilitation of indigenous biodiversity and ecosystem processes must be formulated and tested. The key issue here is the need for scientists to be "prepositional," i.e., not just to present facts about the loss of freshwater biodiversity, but to indicate the research that should be undertaken in support of management strategies designed to reconcile the human use of water with the maintenance of ecosystems and the biodiversity they sustain.

Conclusion

Freshwater biodiversity and freshwater ecosystems are seriously jeopardized by human activities in monsoonal Asia. This is undoubtedly a consequence of the same large and growing human populations that are forcing the Governments in the region to emphasize economic development above everything else. This does not necessarily reflect an absence of legislative frameworks to deal with at least some of the threats to biodiversity. China, for example, has as many environmental laws and policies as some western countries, but they are difficult to enforce and administer because of the many government commissions, departments, and ministries responsible for environmental matters. The State Environmental Protection Administration (SEPA), for instance, must rely on local environmental protection bureaus within municipal and county Governments to enforce regulations (Wu *et al.*, 1999). In this situation, the intentions of state or national bodies and provincial authorities may be at odds, especially when adherence to state regulations on pollution would be costly and could impair local economic development. Management is especially problematic for large rivers such as the Yangtze, which crosses several provincial boundaries and falls within the responsibility of a number of local bureaus and provincial authorities. For this reason, the Fisheries Bureau of the Ministry of Agriculture established the interprovincial Administrative Commission of the Yangtze River Fisheries Resources, which was able to act on a scale that made it possible to introduce an annual fishing moratorium for much of the river. This approach has the support of fisheries scientists in China, although the effectiveness of this authoritarian management system could be enhanced by participatory management, including input from fisher communities (Chen *et al.*, 2004).

Other conflicts that can arise within national borders and between countries are well illustrated by the Mekong River, but the adoption of holistic management as a basis for planning by the Mekong River Commission offers grounds for optimism. There is an urgent need for integrated action and legislation to ensure that critically endangered species, such as the migratory Mekong Giant Catfish, are legally protected in all the countries within their range. Significant obstacles to integrated management of the Mekong Basin will remain as long as China is unwilling to address the downstream consequences of its mainstream dams. The recent suspension of work on the dams along the upper Salween in China could be the first indication of such a change in attitude.

It is clear that additional information could contribute to more effective restoration and management of Asian rivers. However, stakeholder participation and political will are also needed. There are signs of such a commitment in China. A new SEPA Vice-Minister, Pan Yue, has made very forceful use of the media to apply public pressure on recalcitrant ministries and local officials to ensure that legislation and practices that protect the environment are applied. However, the failure of the Ganga Action Plan in India, decades of environmental degradation in China, and the collapse of river fisheries throughout Asia demonstrate that it will be unwise for scientists to assume that Governments and policy makers will institute requirements and practices to protect freshwater biodiversity without societal pressure. Such pressure may be mounting. The activities of non-governmental environmental groups, critical of large-scale dam developments in China, have been widely reported in the state media in recent months. These activities indicate growing societal concern over the management of freshwater resources and parallel longer-established citizen's movements in India, where there is opposition to the construction of the Sardar Sarovar Dam, and in Thailand, where there have been public protests over appropriate operating strategies for the Pak Mun Dam. As part of this wider debate, scientists must communicate the fact that freshwater biodiversity is in crisis and indicate what can be done to ameliorate or improve matters. To be effective, the message must be relevant: books must not simply be added to the library of knowledge; they must be read by those who will apply the information (see Meffe, 2001). An overriding priority is to convey the fact that holistic river management and the restoration of the integrity of riverine ecosystems in Asia will benefit humans through the provision of fisheries resources and clean water, notwithstanding any intrinsic value that may be inherent in genes, species, and natural communities. Successful communication of this message will be an essential first step in halting further impoverishment of biodiversity.

Acknowledgments

This paper is a much-modified version of a presentation given at the Second International Symposium on Riverine Landscapes at Storforsen, Sweden, in August 2004. I am grateful to the SISORL organizers and sponsors for supporting my participation. I also thank Christer Nilsson and three anonymous reviewers for their constructive comments on a draft manuscript, and Lily Ng for assistance with map preparation.

(David Dudgeon, University of Hong Kong.)

Literature Cited

Achard F, H D Eva, H J Stibig, M Mayaux, J Gallego, T Richards and J P Malingreau 2002, Determination of deforestation rates of the world's humid tropical forests. *Science* 297:999-1002.

Arthington A H and B J Pusey 2003, Flow restoration and protection in Australian rivers. *River Research and Applications* 19:377-395.

Arthington A H, J L Rall, M J Kennard and B J Pusey 2003, Environmental flow requirements of fish in Lesotho rivers using the DRIFT methodology. *River Research and Applications* 19:641-666.

Baird I G, B Phylavanh, B Vongsenesouk and K Xaiyamanivong 2001, The ecology and conservation of the smallscale croaker *Boesemania microlepis* (Bleeker 1858-59) in the mainstream Mekong River, southern Laos. Natural History Bulletin of the Siam Society 49:161-176.

Bandyopadhyay J and S Perveen 2004, *The doubtful science of interlinking.* (online) URL: *http://www.indiatogether.org/2004/feb/env-badsci-p1.htm.*

Blaber S J, D A Milton, D T Brewer and J P Salini 2003, Biology, fisheries, and status of tropical shads *Tenualosa* spp. in South and Southeast Asia. *American Fisheries Society Symposium* 35:49-58.

Bradshaw A D 1996, Underlying principles of restoration. *Canadian Journal of Fisheries and Aquatic Sciences* 53 (Supplement 1):3-9.

Braatz S, G Davis, S Shen and C Rees 1992, *Conserving biological diversity: a strategy for protected areas in the Asia-Pacific Region.* World Bank Technical Paper Number 193. Washington, D.C., USA.

Brewer D T, S J Blaber, G Fry, G S Merta and D Efizon 2001, Sawdust ingestion by the tropical shad (*Tenualosa macrura,* Teleostei: Clupeidae): implications for conservation and fisheries. Biological Conservation 97:239-249.

Bunn S E and A H Arthington 2002, Basic principles and ecological consequences of altered flow regimes for aquatic biodiversity. *Environmental Management* 30:492-507.

Chandra R, R K Saxena and R K Tyagi 1990, *Hilsa ilisha* (Ham.) of Ganaga—its glory and downfall, a retrospect. Pages 365-378 in V P Agrawal and P Das, editors. *Recent trends in limnology.* Society of Biosciences, Muzaffarnagar, India.

Chapman E C and D He 1996, Downstream implications of China's dams on the Lancang Jiang (Upper Mekong) and their potential significance for greater regional cooperation, basin-wide. Pages 16-24 in B. Stensholt, editor. *Development dilemmas in the Mekong region.* Monash Asia Institute, Melbourne, Australia.

Chen D, X Duan, S Liu and W Shi 2004, Status and management of the fisheries resources of the Yangtze River. Pages 173-182 in R Welcomme and T Petr, editors. *Proceedings of the Second International Symposium on the Management of Large Rivers for Fisheries.* Volume 1. FAO, Office for Asia and the Pacific, Bangkok, Thailand.

Cui S B 2001, Discussions on several problems of eco-environmental water requirements. *China Water Resources* 8:71-75.

Dudgeon D 1992, Endangered ecosystems: a review of the conservation status of tropical Asian rivers. *Hydrobiologia* 248:167-191.

Dudgeon D 1995, River regulation in southern China: ecological implications, conservation, and environmental management. Regulated Rivers: *Research and Management* 11:35-54.

Dudgeon D 1999, *Tropical Asian streams: zoobenthos, ecology, and conservation.* Hong Kong University Press, Aberdeen, Hong Kong.

Dudgeon D 2000a, Large-scale hydrological alterations in tropical Asia: prospects for riverine biodiversity. *BioScience* 50:793-806.

Dudgeon D 2000b, The ecology of tropical Asian streams in relation to biodiversity conservation. *Annual Review of Ecology and Systematics* 31:239-263.

Dudgeon D 2000c, Riverine wetlands and biodiversity conservation in tropical Asia. Pages 35-60 in B Gopal, W J Junk and J A Davis, editors. *Biodiversity in wetlands: assessment, function, and conservation.* Backhuys Publishers, The Hague, The Netherlands.

Dudgeon D 2000d, Riverine biodiversity in Asia: A challenge for conservation biology. *Hydrobiologia* 418:1-13.

Dudgeon D 2002a, Fisheries: pollution and habitat degradation in tropical Asian rivers. Pages 316-323 in I. Douglas, editor. *Encyclopaedia of global environmental change.* Volume 3. John Wiley,

Dudgeon D 2002b, The most endangered ecosystems in the world? Conservation of riverine biodiversity in Asia. *Verhandlungen Internationale Vereinigung Limnologie* 28:59-68.

Dudgeon D 2003a, The contribution of scientific information to *the conservation and management of freshwater biodiversity in tropical Asia.* Hydrobiologia 500:295-314.

Dudgeon D 2003b, Clinging to the wreckage: unexpected persistence of freshwater biodiversity in a degraded tropical landscape. *Aquatic Conservation: Marine and Freshwater Ecosystems* 13 :93-97.

Dudgeon D, S Choowaew and S C Ho 2000, River conservation in southeast Asia. Pages 279-308 in P J Boon, B R Davies and G E Petts, editors. *Global perspectives on river conservation: science, policy, and practice.* John Wiley, Chichester, UK.

FAO 1995, *Review of the state of world fishery resources: inland capture fisheries.* FAO Fisheries Circular Number 885. FAO, Rome, Italy.

FAO 2000, *The state of world fisheries and aquaculture—2000.* FAO, Rome, Italy.

Federal Interagency Stream Restoration Working Group (FISRWG). 1998. *Stream corridor restoration: principles, processes, and practices.* FISRWG, Springfield, Virginia, USA. (online) URL: *http://www.nrcs.usda.gov/technical/stream_restoration* .

Fu C, J Wu, J Chen, Q Wu and G Lei 2003, Freshwater fish biodiversity in the Yangtze River basin of China: patterns, threats, and conservation. *Biodiversity and Conservation* 12:1649-1685.

Hannah L, D H Lohse, C Hutchinson, J L Carr and A Lankerani 1994, A preliminary inventory of human disturbance of world ecosystems. *Ambio* 23:246-250.

Hassan B 2005, *Fall out of Baglihar Dam.* Dawn Group of Newspapers, Karachi, Pakistan. (online) URL: *http://www.dawn.com/2005/02/14/ebr18.htm.*

Hogan Z, N Pengbun and N van Zalinge 2001, Status and conservation of two endangered fish species, the Mekong Giant Catfish *Pangasianodon gigas and the Giant Carp Catlocarpio siamensis,* in Cambodia's Tonle Sap River. *Natural History Bulletin of the Siam Society* 49:269-282.

Hogan Z, P Moyle, B May, J Vander Zanden and I Baird 2004, The imperiled giants of the Mekong. *American Scientist* 92:228-237.

Hora S L 1942, The effects of dams on the migration of the Hilsa fish in Indian waters. *Current Science* 11:470-471.

International Union for Conservation of Nature and Natural Resources (IUCN). 2004. *The 2004 IUCN Red List of threatened species.* IUCN, Cambridge, UK. (online) URL: *http://www.redlist.org/.*

King J, C Brown and H Sabet 2003, A scenario-based holistic approach to environmental flow assessments for rivers. *River Research and Applications* 19:619-639.

Koehn J D 2004, Carp (Cyprinus carpio) as a powerful invader in Australian waterways. *Freshwater Biology* 49:882-894.

Krishnamurti C R, K S Bilgrami, T M Das and R P Mathur, editors. 1991, *The Ganga: a scientific study.* Northern Book Center, New Delhi, India.

Laurance W F 1999, Reflections on the tropical deforestation crisis. *Biological Conservation* 91 :109-117.

Liao G Z, K X Lu and X Z Xiao 1989, Fisheries resources of the Pearl River and their exploitation. *Canadian Special Publications in Fisheries and Aquatic Science* 106:561-568.

Liu W, S Liu, M Siu and A Y Kuo 2005, Water quality modeling to determine minimum instream flow for fish survival in tidal rivers. *Journal of Environmental Management* 76:293-308.

Low K S 1993, Urban water resources in the humid tropics: an overview of the ASEAN region. Pages 526-534 in M Bonell, M M Hufshmidt and J S Gladwell, editors. *Hydrology and water management in the humid tropics.* UNESCO/ Cambridge University Press, New York, New York, USA.

Marquand R 2005, *China enforcing green laws, suddenly.* Christian *Science* Monitor, Boston, Massachusetts, USA. (online) URL: *http://search.csmonitor.com/2005/0210/p01s02-woap. htm.*

Meffe G K 2001, The context of conservation biology. *Conservation Biology* 15:815-816.

Mekong River Commission (MRC) 2002, *Annual Report 2001.* MRC, Phnom Penh, Cambodia.

Mekong River Commission (MRC) 2005, *Council moves on managing flows.* Mekong News (2005):1.

Nandeesha M C 1994, Fishes of the Mekong River—conservation and need for agriculture. *Naga* 17 :17-18.

Narayana Murty, P. 2000. *The Ganga Action Plan. Scientific Report of the Comptroller and Auditor General of India.* (online) URL: *http://www.cagindia.org/reports/scientific/ 2000_book2/ gangaactionplan.htm.*

Natarajan A V 1989, Environmental impact of Ganga Basin development of gene-pool and fisheries of the Ganga River System. *Canadian Special Publications in Fisheries and Aquatic Science* 106:545-560.

Ni J R, L Jin, Y A Zhao and X Y Liu 2002, Minimum water demand for ecosystem protection in the Lower Yellow River. *Journal of Water Resources* 10:1-7.

Nilsson C and K Berggren 2000, Alterations of riparian ecosystems caused by river regulation. *BioScience* 50:783-792.

Ormerod, S J 2004. A golden age of river restoration *science? Aquatic Conservation: Marine and Freshwater Ecosystems* 14:543-549.

Park Y J Chang, S Leh, W Cao and S Brosse 2003, Conservation strategies for endemic fish species threatened by the Three Gorges Dam. *Conservation Biology* 17:1748-1758.

Payne A I, R Sinha, H R Singh and S Huq 2004, A review of the Ganges Basin: its fish and fisheries. Pages 229-251 in R Welcomme and T Petr, editors. *Proceedings of the Second International Symposium on the Management of Large Rivers for Fisheries.* Volume 1. FAO, Office for Asia and the Pacific, Bangkok, Thailand.

Pearce F 2004, China blamed for Mekong's bizarre flow. *New Scientist.* (online) URL: *http://www.newscientist.com/article.ns?id=dn4819.*

Poff. N L J D Allan, M B Bain, J R Karr, K L Prestegaard, B D Richter, R E Sparks and J C Stromberg 1997, The natural flow regime: a paradigm for river conservation and restoration. *BioScience* 47:769-784.

Poff. N L J D Allan, M A Palmer, D D Hart, B D Richter, A H Arthington, K H Rogers, J L Meyer and J A Stanford 2003, River flows and water wars: emerging *science* for environmental decision-making. *Frontiers in Ecology and the Environment* 1:298-306.

Postel S and B Richter 2003, *Rivers for life: managing water for people and nature.* Island Press, Washington, D.C., USA.

Poulsen A, O Poeu, S Vivarong, U Suntornratana and N Thanh Tung 2002, *Deep pools as dry season fish habitats in the Mekong River Basin.* Mekong River Commission Technical Paper Number 4. MRC, Phnom Penh, Cambodia.

Prakash P 2003, *Linking India's rivers: point and counterpoint.* Indian Institute of Technology, Bombay, India. (online) URL: *http://www.ircc.iitb.ac.in/~webadm/update/archives/August_2003/interlinking1.html.*

Pringle C M 2001, Hydrologic connectivity and the management of biological reserves: a global perspective. *Ecological Applications* 11:981-998.

Rainboth W J 1996, *Fishes of the Cambodian Mekong.* FAO, Rome, Italy.

Richter B D, J V Baumgartner, R Wigington and D P Braun 1997, How much water does a river need? *Freshwater Biology* 37:231-249.

Roberts T R 1993, Just another dammed river? Negative impacts of Pak Mun Dam on the fishes of the Mekong basin. *Natural History Bulletin of the Siam Society* 41:105-133.

Roberts T R 2001a, Killing the Mekong: China's fluvicidal hydropower-cum-navigation development scheme. Natural History Bulletin of the Siam Society 49:143-159.

Roberts T R 2001b, On the river of no returns: Thailand's Pak Mun Dam and its fish ladder. *Natural History Bulletin of the Siam Society* 49 :189-230.

Sala O E, F S Chapin, J J Armesto, R Berlow, J Bloomfield, R Dirzo, E Huber-Sanwald, L F Huenneke, R B Jackson, A Kinzig, R Leemans, D Lodge, H A Mooney, M Oesterheld, N L Poff, M T Sykes, B H Walker, M Walker and D H Wall 2000, Global biodiversity scenarios for the year 2100. *Science* 287:1770-1774.

Shenzhen Daily. 2005. China orders halt to big polluters. (online) URL: *http://www.sznews.com/szdaily/20050208/ca1428140.htm.*

Shi W and G Q Wang 2002, Estimation of ecological water requirements for the Lower Yellow River. *Acta Geographica Sinica* 57:595-602.

State Environmental Protection Administration (SEPA) 1998, *State of the environment, China '97: freshwater environment. SEPA, Beijing, China.* (online) URL: *http://www.zhb.gov.cn/english/SOE/soechina1997/ water/waters1.htm.*

State Environmental Protection Administration (SEPA) 2004, *A report on the state of the environment in China 2003: freshwater environment.* SEPA, Beijing, China. (online) URL: *http://www.zhb.gov.cn/english/SOE/soechina2003/ water.htm.*

Sverdrup-Jensen S 2002, *Fisheries in the Lower Mekong Basin: status and perspectives.* Mekong River Commission Technical Paper Number 6. MRC, Phnom Penh, Cambodia.

Taylor D, P Saksena, P G. Sanderson and K. Kucerra 1999, Environmental change and rain forests on the Sunda shelf of Southeast Asia: drought, fire, and the biological cooling of biodiversity hotspots. *Biodiversity and Conservation* 8 :1159-1177.

Tharme R E 2003, A global perspective on environmental flow assessment: emerging trends in the development and application of environmental flow methodologies for rivers. *River Research and Applications* 19:397-441.

Towards Ecological Recovery and Regional Alliance (TERRA) 2002, Creating catastrophe: China and its dams on the Mekong. *Watershed* 8 (2):42-48.

Usher A D 1996, The race for power in Laos: the Nordic connections. Pages 123-144 in M J G Parnwell and R L Bryant, editors. *Environmental change in South-east Asia: people, politics, and sustainable development.* Routledge, London, UK.

Van Zalinge N, P Degen, C Pongsri, S Nuov, J G Jensen, V H Nguyen and X Choulamany 2004, The Mekong River system. Pages 335-357 in R. Welcomme and T Petr, editors. *Proceedings of the second international symposium on the management of large rivers for fisheries.* Volume 1. FAO Regional Office for Asia and the Pacific, Bangkok, Thailand.

Wan Q, S Fan and Y Li 2003, The loss of diversity in Dabry's Sturgeon (*Acipenser dabryanus* Dumeril) as revealed by DNA fingerprinting. *Aquatic Conservation: Marine and Freshwater Ecosystems* 13:225-231.

Wang H 2003. Biology, population dynamics, and culture of Reeves shad *Tenualosa reevesii. American Fisheries Society Symposium* 35:77-83.

Wang J 1989, Water pollution and water shortage problems in China. *Journal of Applied Ecology* 26 :851-857.

Wei Q F Ke, J Zhang, P Zhuang, J Lou, R Zhou and W Wang 1997, Biology, fisheries, and conservation of sturgeons and paddlefish in China. *Environmental Biology of Fishes* 48:241-255.

Wei Q D He, D Yang, W Zhang and L Li 2004, Status of sturgeon aquaculture and sturgeon trade in China: a review based on two recent nationwide surveys. *Journal of Applied Ichthyology* 20:321-332.

World Commission on Dams (WCD) 2000, *Dams and development: a new framework for decision-making. The report of the World Commission on Dams.* Earthscan Publications, London, UK.

Wu C, C Maurer , Y Wang, S Xue and D L Davis 1999, Water pollution and human health in China. *Environmental Health Perspectives* 107:251-256.

Xinhua News Agency (XNA) 2004, *Fishing ban imposed on Yangtze River.* XNA, Beijing, China. (online) URL: *http://www.china.org.cn/english/environment/84926. htm.*

Xu H, S Wang and D Xue 1999, Biodiversity conservation in China: legislation, plans, and measures. *Biodiversity and Conservation* 8:819-837.

Xu H, D Wang and X Sun 2000, Biodiversity clearing-house mechanism in China: present status and future needs. *Biodiversity and Conservation* 9 :361-378.

Yardley J 2005, Squabbling continues over building of large dam in China: pro-development advocates vs. environmentalists. (online) URL: *http://sfgate.com/cgi-bin/article.cgi?f=/c/a/2005/ 01/03/ MNGVSAKALQ1.DTL.*

Yue P and Y Chen 1998 *China red data book of endangered animals:* Pisces. *Science* Press, Beijing, China.

Zhang S, D Q Wang and Y Zhang 2003, Mitochondrial DNA variation, effective female population size, and population history of the endangered Chinese Sturgeon, *Acipenser sinensis. Conservation Genetics* 4:673-683.

Zheng D Y, J Xia and Y B Huang 2002, Discussion on research of electrical water demand. *International Journal of Hydroelectric Energy* 20 :3-6.

13

Linking Biodiversity Conservation and Livelihoods in India

Kartik Shanker, Ankila Hiremath and Kamal Bawa

This paper speaks of biodiversity conservation in India mainly to curtail the rapid loss of the country's biodiversity. The current activities emphasize interdisciplinary approaches that are specifically used to: generate knowledge that fosters conservation and judicious management of biodiversity; provide the best scientific information to policymakers; design management systems that emphasize decentralization, fairness and equity in the use of resources by civil society; organize and disseminate information for conservation and sustainable use of biodiversity; and train a new generation of leaders to meet current challenge in biodiversity conservation and environmental protection.

In a country like India, millions of people rely on products from natural ecosystems to sustain their livelihood. Natural ecosystems include clean water from watersheds, retention of soil and soil fertility, sequestration of carbon, as well as pollinators and natural predators of pests. For these reasons, rapid, often irreversible, loss of species and ecosystems is of more than just academic concern.

Source: PLOS Biology, November 2005.

Indeed, our understanding of biodiversity in natural ecosystems remains so woefully inadequate that we are unable to fully comprehend the consequences of its loss. With impending climate change and increasing spread of invasive species, the biodiversity crisis is likely to get worse, with far-reaching effects on human societies.

To meet these challenges, we need institutions and scholars that can generate new knowledge, and apply it to resolve our most pressing environmental issues. The Ashoka Trust for Research in Ecology and the Environment (ATREE, *http://www.atree.org*) was established in 1996 to curtail the rapid loss of India's biological resources and natural ecosystems, and to address the environmental, social, and economic dimensions of this decline. We highlight below two examples of ATREE's work from very different ecosystems and regions.

Tribals and Non-Timber Forest Products in the Western Ghats

The Biligiri Rangaswamy Temple Wildlife Sanctuary (BRT), a 540-square-kilometer protected area, forms a part of India's Western Ghats, one of the global hotspots of biodiversity. The area has traditionally been inhabited by an indigenous community, the Soligas, and is also habitat for a number of endangered plants and animals. Soligas have harvested forest products for centuries for their own use and more recently for markets. The interrelated issues of livelihood enhancement of the Soligas and biodiversity conservation have been at the heart of ATREE's work in BRT for close to a decade—along with a partner non-governmental organization, the Vivekananda Girijana Kalyana Kendra; a local community organization, the Soliga Abhivrudhi Sangha; and the Karnataka Forest Department (1). A detailed understanding of the drivers that cause forest loss and degradation is the first step toward its preservation.

We have used demographic models to analyze changes in population structure of nelli *(Phyllanthus emblica and Phyllanthus indofischeri)*, one of the most important NTFPs in BRT. Nelli is an edible fruit high in vitamin C, extracts from which are key ingredients in traditional Indian medicine, and in cosmetics. Our results indicate that population growth rates, on average, are close to rates that would allow full replacement of individuals. Moreover, it is not harvest, per se, but rather the method of harvesting involved (e.g., whether or not it involves the lopping off of branches or cutting of small trees) and the spread of *Loranthus*, a

Figure 1: The Forestland of India's Western Ghats is a Target for Conservation, which must Involve Local Communities

plant parasite that infests mature trees, which affects population growth (2). Management efforts, therefore, need to focus on control of the parasite, and on the use of nondestructive harvest techniques.

Discussion of harvest techniques and parasite removal form part of the annual participatory monitoring meetings with Soliga harvesters. These participatory monitoring meetings also focus on the temporal and spatial patterns in availability of various NTFPs, which are then used to guide harvest decisions. Information collected as part of this participatory monitoring program corroborate the results of scientific monitoring, namely, that populations of nelli, and also those of other important NTFP species, such as the Asian honeybee *(Apis dorsata)*, a source of honey, have remained relatively stable over the last ten years.

It is essential that the benefits accruing to harvesters be maximized for their continued participation in management and conservation. A large part of our effort to strengthen existing institutions has, therefore, been directed toward reform of the Large-scale Adivasi (tribal) Multi-Purpose Society, a government established cooperative society, and a key element in the success of efforts at forest conservation. Non-timber products harvested from the forest can only be sold to the Large-scale Adivasi Multi-Purpose Society. ATREE is also working with Soliga farmers to increase agricultural productivity, enhance on-farm diversity, and improve soil and water conservation. It is hoped that through simple agricultural interventions we can achieve a greater on-farm contribution toward subsistence and cash needs, thereby reducing the extent of dependence on NTFP.

BRT is the only forest area in India where production and extraction of NTFPs are being monitored, and where the local community is involved in such monitoring. In a recent meeting with the Forest Department, a committee comprising members of the Soliga community, Vivekananda Girijana Kalyana Kendra, and ATREE was proposed to provide suggestions to the Forest Department on management of the protected area. If formalized, this would make BRT the first protected area to have such a three-way collaboration between managers, the local community, and researchers, and would be a model for other protected areas in the country.

Protecting Turtles with Fisherfolk

On the other side of the country, on the coast of Orissa, are the nesting grounds of olive ridley sea turtles *(Lepidochelys olivacea).* This is one of three rookeries worldwide where arribadas, the synchronous mass nesting of thousands of ridley turtles, occurs. Genetic studies demonstrate that this is a unique population, which may be ancestral to olive ridleys in other ocean basins (3). In the last decade, however, 10,000 turtles have been counted dead on the Orissa coast each year due to fishery-related activities. And many more are likely to have died, since not all turtles killed in nets are washed ashore (4).

Research indicates that olive ridleys remain in small offshore congregations during the breeding season, and are not diffusely distributed along the entire coast. Discrete reproductive patches off the mass nesting beaches are usually not more than 50 square kilometers; however, the location of these patches may vary over time (5). Thus, the creation of sanctuaries or protected areas with fixed boundaries may not be effective. Instead, conservationists have focused on the enforcement of laws, including existing fishery regulations such as the 1983 Orissa Marine Fisheries Regulation Act, which stipulates that mechanized fishing is prohibited within five or ten kilometers of the coast, depending on boat size.

Despite the investment of large amounts of effort and funds by the government and civil society groups to patrol nearshore waters, trawlers continue to fish illegally in nearshore waters, causing continuing mortality of olive ridleys. Rather, the anti-trawler programs, coupled with the media coverage, have severely polarized fishing communities and conservation groups. Even traditional fish-worker associations in Orissa joined in protests with the trawler owners, since they perceived turtle conservation as being anti-people, even though most of the Orissa Marine Fisheries Regulation Act regulations were designed to protect traditional fishing rights rather than turtles.

In fact, if fishery laws had been enforced for the reasons that they were originally instituted, namely, to protect traditional fisherfolk and their livelihoods, it is likely that their implementation would have received far wider support. And as a result, sea turtles would then have been protected from mechanized fishing. To achieve this goal, the Coastal and Marine Programme at ATREE created and facilitated a common platform for sea turtle conservation in Orissa. In December 2004, ATREE organized a meeting in Bhubaneshwar that was attended by key fish-worker organizations (Orissa Traditional Fish Workers' Union and United Artists Association), local community organizations, and non-governmental organizations such as Project Swarajya, Wildlife Society of Orissa, Worldwide Fund for Nature, Greenpeace, and others. The group named itself the Orissa Marine Resources Conservation Consortium, and has been working together to achieve common marine conservation goals.

The Orissa Marine Resources Conservation Consortium has held numerous follow-up meetings in 2005. The activities identified to meet these common goals included meetings for fisherfolk representatives and other stakeholders on fisheries management and turtle conservation legislation in April, 2005. ATREE has produced illustrative local language booklets on fisheries conservation and turtle protection measures in Orissa to help local communities understand their rights and regulations. Booklets, along with other visual aids such as posters illustrating fishing regulations, have been distributed to various stakeholders at the mass nesting beaches. The Orissa Marine Resources Conservation Consortium plans to promote community-based marine conservation practices and appropriate environmentally sustainable coastal development, addressing the issues of marine biodiversity and resource use.

The Big Picture

With its headquarters in Bangalore, ATREE also has offices in Delhi, and northeast India, where its programs are directed toward conservation of the eastern Himalayan region. ATREE's current activities are grouped under research and action, education, and outreach. Specfically, ATREE uses interdisciplinary approaches to (1) generate knowledge that fosters conservation and judicious management of biodiversity, (2) provide the best scientific information to policymakers, (3) design management systems that emphasize decentralization, fairness, and equity in the use of resources by civil society, (4) organize and disseminate information for conservation and sustainable use of biodiversity, and (5) train a new generation of leaders to meet current challenges in biodiversity conservation and environmental protection. The various activities are organized under three centers: the Center for Conservation Science; the Center for Conservation, Governance, and Policy; and the Center for Ecoinformatics *(http:www.ecoinfoindia.org)*. Most of the existing staff work in diverse areas of natural and social sciences including biodiversity characterization, forest ecology, taxonomy, conservation genetics, landscape ecology, hydrology, environmental sociology, and ecological economics, in the Center for Conservation Science. All three centers are, or would be, engaged in education and outreach activities designed to build the capacity of academic, governmental, and non-governmental organizations to meet the growing list of contemporary environmental challenges.

(Kartik Shanker, Ankila Hiremath, and Kamal Bawa, Ashoka Trust for Research in Ecology and the Environment, Bangalore, India. They can be contacted at kartik@atree.org).

References

1. Bawa KS (1999), Prescriptions for conservation. In: Biodiversity Conservation Network. Final stories from the field. Washington (D.C.): Biodiversity Support Program. pp. 48–55.

2. Sinha A, Bawa KS (2002), Harvesting techniques, hemiparasites and fruit production in two non-timber forest tree species in south India. For Ecol Manage 168: 289–300.

3. Shanker K, Pandav B, Choudhury BC (2004), An assessment of the olive ridley turtle *(Lepidochelys olivacea)* nesting population in Orissa, India. Biol Conserv 115: 149–160.

4. Shanker K, Ramadevi J, Choudhury BC, Singh L, Aggarwal RK (2004) Phylogeography of olive ridley turtles *(Lepidochelys olivacea)* on the east coast of India: Implications for conservation theory. Mol Ecol 13: 1899–1909.

5. Pandav B (2000), Conservation and management of olive ridley sea turtles on the Orissa coast (thesis). Bhubaneshwar (India): Utkal University. 106 p. Available from Wildlife Institute of India, Dehradun, India.

Biodiversity Profile of India

India contains a wealth of biodiversity in its forests, wetlands and marine areas. It has many institutes and university departments, wherein a large number of scientists and technicians have been engaged in inventory, research, and monitoring. The general state of knowledge about the distribution and richness of the country's biological resources is therefore fairly good.

The current state of knowledge suggests that the country has many endemic plant and vertebrate species. Among plants, species endemism is estimated at 33% by Botanical Survey of India in 1983. North-east India, the Western Ghats and the north-western and eastern Himalayas are the most endemic. The Andaman and Nicobar Islands reportedly contribute at least 220 species to the endemic flora of India.

Four endemic mammalian species of conservation significance occur in the Western Ghats. They are the Lion-tailed macaque Macaca silenus, Nilgiri leaf monkey Trachypithecus johni (locally better known as Nilgiri langur Presbytis johnii), Brown palm civet Paradoxurus jerdoni and Nilgiri tahr Hemitragus hylocrius. Endemic bird species number over fifty, with distributions concentrated in areas of high rainfall in the country. Endemism in the Indian reptilian and amphibian fauna is quite high with around 187 endemic reptiles, and 110 endemic amphibian species. Eight amphibian genera are not found outside India. They include, among the caecilians, Indotyphlus, Gegeneophis and Uraeotyphlus; and among the anurans, the toad Bufoides, the microhylid Melanobatrachus, and the frogs Ranixalus, Nannobatrachus and Nyctibatrachus. Perhaps most notable among the endemic amphibian genera is the monotypic Melanobatrachus which has a single species known only from a few specimens collected in the Anaimalai Hills in the 1870s (Groombridge, 1983). It is possibly most closely related to two relict genera found in the mountains of eastern Tanzania.

Over and above, the country contains 172 species of animals considered globally threatened by IUCN, or 2.9% of the world's total number of threatened species (Groombridge, 1993). These include 53 species of mammal, 69 birds, 23 reptiles and 3 amphibians. India contains globally important populations of some of Asia's rarest animals, such as the Bengal Fox, Asiatic Cheetah, Marbled Cat, Asiatic Lion, Indian Elephant, Asiatic Wild Ass, Indian Rhinoceros, Markhor, Gaur, Wild Asiatic Water Buffalo etc.

Sources: Botanical Survey of India (1983), Flora and Vegetation of India – An Outline. Botanical Survey of India, Howrah. 24 pp

Groombridge B (1983), Comments on the rain forests of southwest India and their herpetofauna. Paper prepared for the Centenary Seminar of the Bombay Natural History Society, 6-10 December, 1983. 18 pp. Revised, January 1984.

Groombridge B (ed) 1993, The 1994 IUCN Red List of Threatened Animals. IUCN, Gland, Switzerland and Cambridge, UK. lvi + 286 pp. http://ces.iisc.ernet.in/hpg/cesmg/indiabio.html

Biological Diversity Act, 2002

The Biological Diversity Act, 2002 of India is an Instrument for providing avenues for effective conservation of biological diversity, sustainable use of its components and fair and equitable sharing of the benefits arising out of the use of biological resources, knowledge and for matters connected therewith or incidental thereto. This Act is meant to achieve the following objectives:

- The conservation of biodiversity;
- The sustainable use of biological resources; and
- Equity in sharing benefits from such use of resources.

Some key provisions aimed at achieving the above objectives are:

1. Prohibition on transfer of Indian genetic material outside the country, without specific approval of the Indian Government;
2. Prohibition on anyone claiming an Intellectual Property Rights (IPR), such as a patent, over biodiversity or related knowledge, without permission of the Indian Government;
3. Regulation of collection and use of biodiversity by Indian nationals, while exempting local communities from such restrictions;
4. Measures for sharing of benefits from the use of biodiversity, including transfer of technology, monetary returns, joint Research & Development, joint IPR ownership, etc.;
5. Measures to conserve and sustainably use biological resources, including habitat and species protection, Environmental Impact Assessments (EIAs) of projects, integration of biodiversity into the plans, programmes, and policies of various departments/sectors;
6. Provisions for local communities to have a say in the use of their resources and knowledge, and to charge fees for this;
7. Protection of indigenous or traditional knowledge, through appropriate laws or other measures such as registration of such knowledge;
8. Regulation of the use of genetically modified organisms;
9. Setting up of National, State, and Local Biodiversity Funds, to be used to support conservation and benefit-sharing; and
10. Setting up of Biodiversity Management Committees (BMC) at local village level, State Biodiversity Boards (SBB) at state level, and a National Biodiversity Authority (NBA).

Some organizations, however argue that the basic framework of the Act is problematic, since it accepts intellectual property rights on biodiversity, could be used to further commercialize biodiversity, and does not truly empower communities. Others feel that the Act provides some potential for checking bio-piracy, achieving conservation, and facilitating community action. They stress that a combination of strong rules, and amendments related to the above points, would help strengthen this potential.

Sources: http://www.genecampaign.org/home/Biological%20Diversity%20Act%202002.pdf http://www.kalpavriksh.org/kalpavriksh/f1/f1.1/bdbdcamp/Biodiversity%20Act%20and%0Rules,%20basic%20note,%20Final,%2030.9.2004.doc

14

Monitoring Biodiversity of Select Restoration Sites in New Zealand

Jesse Bishop, Russell G Congalton and Mimi L Becker

Much of New Zealand is now covered by non-native plants and animals. In recent years there has been increased effort to mitigate these effects through a process of ecosystem restoration. The GLOBE Land Cover/Biology Investigation Team at the University of New Hampshire, working collaboratively with GLOBE New Zealand (Dr. John Lockley) and Landcare Crown Research Institute (Dr. Daniel Rutledge) at the University of Waikato, Hamilton, New Zealand, will use GLOBE student collected land cover data to create land cover maps and perform change analysis in order to monitor land cover changes at selected ecosystem restoration sites in New Zealand. A crosswalk between GLOBE's Modified UNESCO Classification (MUC) system and the Land Environments of New Zealand (LENZ) system will be developed to allow for the effective sharing of data between these projects.

Source: www.globe.gov, Reprinted with permission.

Introduction

The New Zealand Government has recognized and taken steps to halt the loss of native biodiversity within the country. Initially, restoration efforts were focused on offshore islands. Current efforts extend to mainland islands, restoration sites within a larger landscape that are isolated by a natural or constructed barrier. Student data will be collected in and around these sites and used as part of a larger data set to develop thematic land cover maps. Numerous pre-processing, data exploration, classification and change detection techniques will be used in this project. The three main outputs from this project will be the thematic land cover maps, change detection information, and a crosswalk between the Modified UNESCO Classification (MUC) system and the Land Environments of New Zealand (LENZ) system.

New Zealand

As a result of 80 million years of isolated evolution, a majority of the flora and fauna found in New Zealand are unique in the world. Over 80% of the native vascular plants in New Zealand are endemic. Although New Zealand was one of the last major land areas to be impacted by human settlers, these impacts have been intense. By the late 20th century, more than 60% of the land has been converted from native vegetation through human activities, including development, farming and exotic forestry. Much of the remaining native vegetation exists in isolated pockets in the mountains and other areas that have been inaccessible or are not economically feasible to develop. The result can be seen in the New Zealand Biodiversity Index, a measurement developed by the New Zealand Department of Conservation that shows a steady decline in biodiversity since the time of settlement (New Zealand Department of Conservation, 2000).

Human settlers carried with them many species of plants and animals to New Zealand. These exotic species have had a dramatic influence on the landscape. Introduced species, both in the wild and cultivated, outnumber native and naturalized species. Native wildlife is affected by this shifting land use. Many species are unable to survive in these modified habitats, leading to small, fragmented populations and widespread extinctions (New Zealand Department of Conservation, 2000).

The New Zealand Government has recognized both the intrinsic and economic values of biodiversity. The native flora and fauna represent the unique characteristics of New Zealand. The kiwi, a flightless bird, and the silver fern are national icons, and both are being pushed toward extinction by current land uses. The Maori, the indigenous people of New Zealand, believe humans have a common ancestry with animals. They are very concerned with the loss of biodiversity (New Zealand Department of Conservation, 2000).

Tourism is an important part of the economy of New Zealand. The distinctiveness of the natural environment and New Zealand's clean and green image are the major draw for tourists. Restoration of native ecosystems is necessary to maintain this segment of the economy. Patterson and Cole (1999) have estimated the total value of New Zealand's indigenous biodiversity, including direct economic benefits and intrinsic values, to be twice the New Zealand Gross Domestic Product. The Government of New Zealand has developed The New Zealand Biodiversity Strategy to address these concerns. Many branches of Government are implementing the strategy. One important aspect of this strategy is to integrate biodiversity considerations into the National Strategy for Education (New Zealand Department of Conservation, 2000). This collaborative project between GLOBE New Zealand and the Land Cover/Biology Team will expose students to the techniques that can be used to monitor biodiversity.

Remote Sensing and Land Cover Mapping

Remote sensing provides an opportunity for large-scale measurements. Ground-based reference data increases the usefulness of these data (Botkin and Estes, 1984). Data gathered by remote sensing can be analyzed to develop land cover classification maps and detect land cover change (Green *et al.*, 1994). Accurate reference data are necessary to validate the interpretation and mapping of remotely sensed land cover data but has not always been collected (Becker *et al.*, 1998).

Land Environments of New Zealand (LENZ), launched in June 2003, is a quantitatively-based landscape classification scheme that aims to assist biodiversity conservation and natural resource management. This classification relies on the relationships between species and their environment in order to group landscapes with similar ecological characteristics, not just land cover. One of many advantages of this system is the ability to predict prehistoric land cover based on existing and historical characteristics of disturbed sites (Landcare Research *http://lenz.landcareresearch.co.nz/*).

Globe

The GLOBE Program was introduced to New Zealand in 2000. The first schools were trained in the beginning of 2001 (Lockley, 2002). At present, there are over 100 schools involved with the GLOBE program throughout New Zealand. The GLOBE Program in New Zealand is an important part of the growing area of environmental education (Lockley, 2002).

One goal of the Land Cover/Biology Team at the University of New Hampshire is to use student data to validate thematic land cover maps created from satellite images (Fried, 1997). In order to do this, the Land Cover/Biology Team first had to develop a suitable classification scheme. The MUC (Modified UNESCO Classification) system was developed by adding two developed classes to the existing UNESCO classification system (Becker *et al.*, 1998; The GLOBE Program 2000). The MUC system was designed to be suitable for any location globally. Then, the Land Cover/Biology Team developed standardized collection protocols that instruct the students how to collect land cover and biology data (Becker *et al.*, 1998; Rowe, 2001).

Data Collection, Accuracy Assessment and Land Cover Mapping

Rock and Lauten (1996) have shown that data collected by students as part of the *Earth Day: Forest Watch* program at the University of New Hampshire may be reliable and accurate. Furthermore, they have found that the data can be a great contribution to research by increasing the sample population and expanding the study areas. Budd (1997) and Rowe (2001) have shown that data collected by GLOBE students in the United States have an acceptable level of accuracy.

Reference data must be used to check the accuracy of a classified map (Plourde and Congalton, 2003). The most common method of assessing accuracy of a land cover classification map is to create an error matrix (Congalton, 1991; Plourde and Congalton, 2003). In order for this assessment to be valid, it is necessary for the reference data to be highly accurate (Congalton, 1991).

Chuvieco and Congalton (1988) have described three phases for computerized classification of remotely sensed data: training, assignment of non-sampled pixels, and output and assessment. Training may be the most important step in the process. Significant error can result from bad training, biasing the results.

Change Detection

Digital change detection attempts to measure differences in land cover between two points in time by focusing on certain variables of interest (i.e., spectral reflectance of land cover) and controlling other unwanted variation (i.e., atmospheric conditions, sensor differences) (Green *et al.*, 1994). There are many different techniques used for multi-date digital change detection. Change detection can be performed on imagery acquired from airborne sensors or on thematic maps created from those images. Each technique has associated advantages and disadvantages. Image-to-image change detection techniques generally require less effort and provide change/no change information. Map-to-map change detection techniques require more effort from the analyst but provide detailed change-from/change-to information. Map to map change detection accuracy is highly dependent on the skill of the analyst (Jensen, 1996).

Approach and Methods

- Develop contacts with New Zealand teachers interested in the GLOBE Program.
- Train New Zealand students in GLOBE protocols.
- Assist New Zealand students conducting data collection.
- Collect reference data.
- Create thematic land cover maps.
- Perform a digital change detection for each site to analyze "from-to" change.
- Develop a crosswalk between the LENZ and MUC classification systems.

Study Areas

There are a number of sites undergoing restoration in New Zealand. This project will focus on four of those sites, Maungatautari, the Karori Wildlife Sanctuary,

Bushy Park and Rotoiti. These sites were selected because of their size, accessibility, and their proximity to GLOBE schools. Landsat imagery from 1990, 2000, and 2002 is available for three of the sites. Currently, imagery containing the Bushy Park site is only available from 2002.

Maungatautari is a volcanic dome that rises alongside the Waikato River, surrounded by farmland of the central plain of the North Island of New Zealand. A 3,400-hectare native forest covers the mountaintop. Construction of a 47 km pest-proof fence that will eventually surround the forested peak has begun. Once the fence is completed, all warm-blooded animal pests will be removed (Maungatautari Ecological Island Trust, *http://www.maungatrust.org*).

The Karori Wildlife Sanctuary is a 252-hectare native forest 2 km from Wellington, the capital of New Zealand. An 8.6 km pest-proof fence surrounds the sanctuary. Restoration efforts are in progress. About 50% of the sanctuary will be restored; the rest will be allowed to develop naturally (Karori Wildlife Sanctuary, *http://www.sanctuary.org.nz*).

Bushy Park is a small (88 hectare) native forest in Wanganui on North Island. The forest is part of a homestead donation. The wetland forest is a major attraction for guests staying at the homestead (Bushy Park Homestead, *http://www.bushypark-homestead.co.nz*).

The Rotoiti site is located in the Nelson Lakes region of the northern South Island. Restoration efforts focusing on 825 hectares of native southern beech (Nothofagus sp) forest on the shore of Lake Rotoiti began in 1997. This area is managed as a mainland island and extensive work has been conducted to eradicate non-native species (New Zealand Department of Conservation, *http://www.doc.govt.nz*).

Tools

The most important tool for data collection on the GLOBE Land Cover Sample Sites is the *GLOBE Teacher's Guide* (The GLOBE Program, 2003). The guide contains all of the protocols and data sheets for collecting data. It also contains instructions and templates for constructing low-cost clinometers and densiometers. In order to collect the data, students will need a Landsat TM image showing their study area, a MUC Field *Guide or MUC System Table and MUC Glossary of Terms*, a GPS receiver, a camera, a compass, measuring tapes, vegetation field guides, grass clippers, a densiometer, and a clinometer. The equipment will be supplied by the GLOBE schools.

The students will create a land cover map using MultiSpec software, a freeware image processing package. The UNH Team will create land cover maps and perform change detection using both MultiSpec and ERDAS Imagine 8.6 along with ESRI's suite of GIS tools.

Note

This material is based upon work supported by the National Science Foundation under grant No. Go-022375. Any opinions, Findings and Conclusions or recommendations expressed in the material are those of the authors and do not necessarily reflect the views of National Science Foundation.

(Jesse Bishop, Russell G Congalton, and Mimi L Becker, Department of Natural Resources, 215 James Hall University of New Hampshire Durham, NH. They can be reached at jesse.bishop@mac.com, russ.congalton@unh.edu, mlbecker@cisunix.unh.edu respectively.)

Literature Cited

Becker M L, R G Congalton, R Budd and A Fried 1998, A GLOBE Collaboration to develop land cover data collection and analysis protocols. *Journal of Science Education and Technology,* 7(1):85-96.

Botkin D B and J E Estes 1984, Studying the earth's vegetation from space. *BioScience,* 34(8):508-514.

Budd R J 1997, *The Accuracy of Land Cover Data Collected by Students (Grade 1-12) for the Global Learning to Benefit the Environment* (GLOBE) Program. M S thesis, University of New Hampshire, Durham, New Hampshire, 275 p.

Chuvieco E and R G Congalton, 1988, Using cluster analysis to improve the selection of training statistics in classifying remotely sensed data. *Photogrammetric Engineering & Remote Sensing,* 54(9):1275-1281.

Congalton R G 1991, A review of assessing the accuracy of classification of remotely sensed data. Remote Sensing of Environment, 37:35-46.

Fried A A 1997, *Training Material Design Criteria for the Implementation of Student-Teacher-Scientist Partnerships in the Elementary and Middle School Levels:* with a focus on the GLOBE Program, M S thesis, University of New Hampshire, Durham, New Hampshire, 187 p.

Green K D Kempka and L Lackey 1994, Using remote sensing to detect and monitor land-cover and land-use change. *Photogrammetric Engineering & Remote Sensing,* 60(3): 331-337.

Jensen J R 1996, *Introductory Digital Image Processing:* A Remote Sensing Perspective (K.C. Clarke, editor), Prentice Hall, Upper Saddle River, N.J., 318 p.

Lockley J 2002, Country Report: New Zealand. The 7th Annual GLOBE Conference, 22-26 July 2002, Chicago, Illinois, USA.

New Zealand Department of Conservation 2000, New Zealand Biodiversity Strategy, URL: *http://www.biodiversity.govt.nz/picture/doing/nzbs/index.html,* New Zealand Department of Conservation, Wellington, New Zealand, (date last accessed: 15 October 2003).

Patterson M and A Cole 1999, *Assessing the Value of New Zealand's* Biodiversity. Occasional Paper Number 1. School of Resource and Environmental Planning, Massey University, Auckland, New Zealand.

Plourde L and R G Congalton 2003, Sampling method and sample placement: how do they affect the accuracy of remotely sensed maps? *Photogrammetric Engineering & Remote Sensing,* 69(3):289-297.

Rock B N and G N Lauten 1996, K-12th Grade Students as Active Contributors to Research Investigations. *Journal of Science Education and Technology,* 5(4):255-266.

Rowe R 2001, *Land Cover Classification of Remotely Sensed Data: Assessing the Accuracy Using Student-Collected Reference Data from the Global Learning and Observations to Benefit the Environment (GLOBE) Program,* M.S. thesis, University of New Hampshire, Durham, New Hampshire, 125 p.

The GLOBE Program, 2000, *MUC Field Guide,* GLOBE & USGPO, Washington, D.C., 117 p.

The GLOBE Program, 2003. *The GLOBE Teacher's Guide,* GLOBE & USGPO, Washington, D.C., 1936 p.

Vogelmann J E and B N Rock 1988, Assessing forest damage in high-elevation coniferous forests in Vermont and New Hampshire using Thematic Mapper data. *Remote Sensing of Environment,* 24:227-246.

Biodiversity in Australia

Australia has an enviably rich biodiversity and as a 'developed' nation, it has a special responsibility for conserving as well as managing its biodiversity effectively. It is estimated that there are 13.6 million species of plants, animals and micro-organisms on earth. Australia has about one million of these, which represents more than 7 percent of the world's total and is more than twice the number of species in Europe and North America combined. Megadiversity describes countries with very high levels of biodiversity. Twelve of the megadiverse countries, including Australia, contain about 75% of earth's total biodiversity. Other megadiverse countries include Brazil, Colombia, Ecuador, Peru, Mexico, Democratic Republic of the Congo, Madagascar, China, India, Indonesia and Malaysia.

Australia's high number of endemic species is largely a result of its long period of separation from other continents. Australia was once part of the great southern super-continent Gondwana, which also included South America, Africa, India and Antarctica. Gondwana began to break up some 140 million years ago and about 50 million years ago, Australia eventually split from Antarctica. Australia's unique biodiversity is mostly explained by the isolation of our continent from other land masses.

The Great Barrier Reef is an Australian icon, symbolic of the wealth of Australia's biological diversity. It is the nursery to fishing industries and is visited by thousands of tourists each year, bringing hundreds of millions of dollars into the country.

Among the most recent innovations is a project run by farmers, community leaders and researchers in South Australia, which is showing how conserving native grasses can boost farmers' productivity. In 1999, the South Australian Department for Environment and Heritage received $683,490 funding from the Australian Government's Natural Heritage Trust to demonstrate that appropriate grazing management can both allow native pastures to be grazed for production and result in improved conservation of native grasslands. The project was administered by a community group, the Mid North Grasslands Working Group. Since 2003 the Group has been supported by further funding from the Land, Water & Wool program run by Land & Water Australia, the Natural Heritage Trust, and the Northern and Yorke Agricultural District Integrated Natural Resource Management Committee.

Sources: http://www.environment.gov.au/biodiversity/publications/greenhouse/pubs/greenhouse.pdf

http://www.amonline.net.au/biodiversity/what/australia.htm

http://www.nrm.gov.au/publications/factsheets/pubs/bio-saving-grasses.pdf

15

Biodiversity Conservation, Sustainable Development, and the US Man and the Biosphere Program: Past Contributions and Future Directions

P N Manley and D C Hayes

US Man and the Biosphere (MAB) Program is part of the United Nations Educational, Scientific, and Cultural Organization (UNESCO) MAB program, and is one of six regional MAB programs that span the globe. The MAB Program was created in 1971 with the goal to explore, demonstrate, promote, and encourage harmonious relationships between people and their environments. Biosphere reserve networks are a primary vehicle for accomplishing MAB goals and serve four basic functions: 1) conserve biodiversity, 2) demonstrate sustainable development approaches, 3) support research and monitoring related to local, national, and global issues of conservation and sustainable development, and 4) build social capacity for sustainable development through education and training. The US network, established in 1976, consists of 47 biosphere

Source: www.fs.fed.us, Reprinted with permission.

reserves that represent a diversity of ecosystems. The US MAB program is in a period of reflection and revitalization as it nears 30 years of commitment and contribution to biological diversity conservation and sustainable development. Major accomplishments of the US MAB program have been through many different institutions, such as the Information Center for the Environment's (ICE) development of a biodiversity database that serves to document species occurrences for protected areas around the world (www.ice.ucdavis.edu), the development of the monitoring and assessment of biodiversity program research and education activities by the Smithsonian Institution (www.nationalzoo.si.edu/conservation and science/MAB), and Southern Appalachian MAB program's exemplary achievements in promoting environmental health and stewardship in natural and cultural ecosystems in the Southern Appalachian mountains. Future emphasis areas for the US MAB program include refreshed operating principles based on a model of integrated human and natural ecosystems, enhanced networking capabilities among reserves within the US and around the world, innovative advances in global change monitoring, sustainability research, and education at US biosphere reserves, and accelerated development of solutions to key challenges to sustainability.

Introduction

Sustainability is increasingly a focus of national and international attention in terms of dialogue, monitoring, research, development, and policy (Sayer and Campbell, 2003). However, recognition of the importance of sustainability is not new – indeed, it is as old as human societies themselves (Diamond, 1999). Over the past 50 years, many advances have been made in our understanding of the limitations and sensitivities of the biosphere, and in our use of technology to minimize detrimental effects of human activities on the environment. The United Nations Educational, Scientific, and Cultural Organization's (UNESCO) Man and the Biosphere (MAB) program, founded over 30 years ago, is an example of a significant international effort to enhance and promote sustainability, and it has made significant contributions toward this goal, both in the US and throughout many countries around the world. Much has changed since the 1970s, most significantly being that human populations have increased by approximately 50 percent, from 4 billion to 6 billon people globally, and similarly from 200 million to 300 million in the United States (US Census Bureau, 2004). As we begin a new millennium, it seems prudent to take stock of US MAB's past accomplishments, operational short-comings, current and future opportunities, and pressing priorities to sharpen its focus and continue its insightful offerings to tomorrow's sustainability challenges.

MAB Program Fundamentals

The MAB Program was launched in 1971 to facilitate intergovernmental cooperation in promoting harmonious relationships between people and their environments. As such, MAB was the first deliberate international initiative to work toward sustainable development. Specifically, the goals of the MAB program are to: (1) foster the rational use and conservation of the resources of the biosphere and the improvement of the global relationship between man and the environment; and (2) to predict the consequences of today's actions on tomorrow's world and thereby increase man's ability to manage efficiently the natural resources of the biosphere. US MAB Program is one of six regional MAB programs in 97 countries that span the globe (UNESCO, 2003). The United States was one of the first countries to establish a national MAB organization and begin establishing biosphere reserves as part of a national network. The United States is a member of the regional EuroMAB program, along with 30 other countries in Europe and North America. The objective of the US MAB program mirrors those of the international MAB program with an added emphasis on a balance between social and ecological systems, "To demonstrate and advance a sustainable balance between conserving biological diversity and promoting human development, while maintaining associated cultural values" *(www.euromab.org/general_information/geninfo.html).*

The US MAB program consists of three primary components: a National Committee, Research Directorates, and a Biosphere Reserve Network. The National Committee is comprised of a diversity of entities, including federal agencies, academic institutions, and non-governmental institutions. The National Committee provides national direction for all aspects of the program, such as the establishment of research directorates, allocation of funds and resources to biosphere reserves, international relations, and fund raising. Research directorates are pivotal positions designated to promote focused research and education to further our understanding of ecosystem sustainability and speed the development, availability, and application of key information, useful tools, and effective practices. Five research directorates were established: temperate ecosystems, high latitude ecosystems, marine and coastal ecosystems, tropical ecosystems, and human-dominated ecosystems.

Biosphere reserve networks are the primary landbased vehicle for accomplishing MAB objectives and serve three basic functions: 1) conservation – contribute to the conservation of landscapes, ecosystems, species, and genetic variation; 2) development – foster economic and human development which is socio-culturally and ecologically sustainable; and 3) logistic support – support for demonstration projects, environmental education and training, research and monitoring related to local, regional and global issues of conservation and sustainable development (UNESCO, 1995). UNESCO's criteria for establishing a biosphere reserves are as follows: (1) contain a mosaic of ecological systems representative of a major biogeographic region, including a gradation of human influences; (2) contain areas significant for biodiversity conservation; (3) offer opportunities to explore and demonstrate approaches to sustainable development on a regional scale; 4) extend over an appropriately sized area to serve the first three functions; 5) identify appropriate zones (core, buffer and transition) to accomplish functions; 6) establish an organizational structure that provides for the involvement and participation of a range of authorities, communities, and interests; and 7) plans, programs, and policies that support the functions and activities of the biosphere reserve (UNESCO, 1995). Biosphere reserves are intended to have three zones delineated: (1) one or more core areas, consisting of legally protected area(s) managed to sustain indigenous biota and natural processes; (2) a buffer zone, consisting of a legally or administratively established area that typically adjoins or surrounds the core area; and (3) a transition area, which surrounds the core area and buffer zone and supports a variety of resource uses and human activities characteristic of the larger region (Figure 1).

Figure 1: Three Management Zones Associated with Biosphere Reserves (from UNESCO)

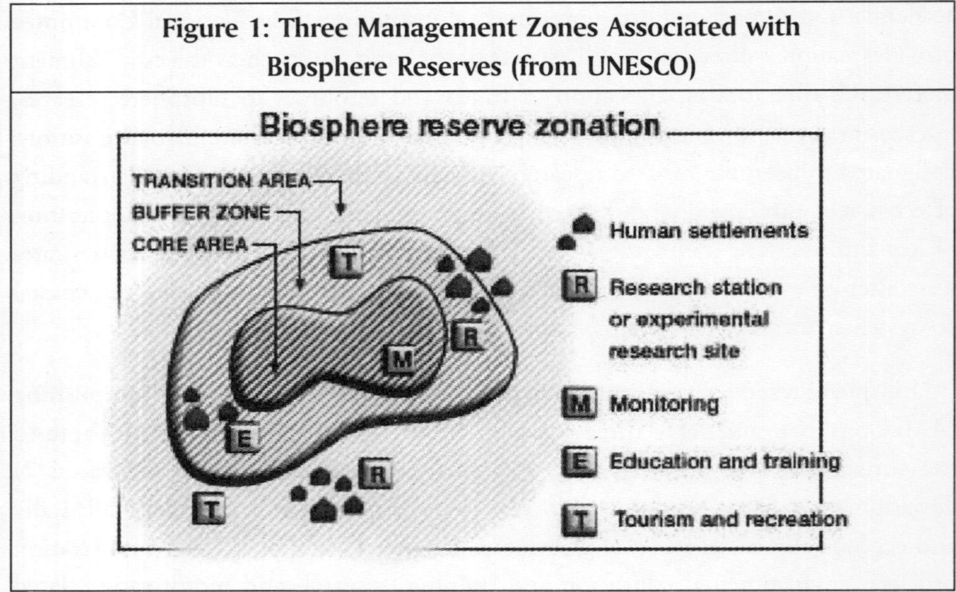

Highlights of US MAB Contributions

Overall, UNESCO views MAB as instrumental in helping to reconciling conflicts between conservation and resource use by contributing to a more in-depth understanding of social-ecological interactions, and developing tools and techniques that can be used to implement ecosystem approaches promoted by Convention on Biological Diversity *(www.biodiv.org)*. The US MAB program has made significant contributions to these ends in conserving biological diversity and promoting sustainable development in the US and around the globe. Its contributions have taken a multitude of forms and have been accomplished through many different institutions.

Throughout the tenure of the US MAB program, individual biosphere reserves have made a plethora of research, inventory, monitoring, and education contributions to local, regional, and international communities. In 1995, stellar contributions of 12 biosphere reserves were highlighted in US MAB publication (Anonymous, 1995). The cornucopia of activities included: A symposium on bioregional biodiversity sustainability; coordinated research between a US and French coastal biosphere reserve on ecological sustainability issues; a regional workshop to introduce local educators to the objectives and activities of biosphere reserve programs; a cooperative research program to accomplish data compilation, analysis, and modeling for an ecoregion; and a multi-agency assessment of water quality within a large-scale community watershed.

In the 1980s, the Smithsonian Institution, as a partner in the US MAB program, developed their Monitoring and Assessment of Biodiversity program *(www.nationalzoo.si.edu/conservationandscience/MAB).* The mission of the program is to enable the implementation of biodiversity monitoring projects and to inform adaptive management planning around the world. Over its nearly 20 years of operation, this program has made substantial advances in research, development, and education in the arenas of biodiversity inventory and monitoring. For example, they offer formal educational opportunities in the form of intensive courses in biodiversity assessment and monitoring, and environmental leadership. They also have an active research program in over 10 locations around the world that strives to accomplish four main objectives: 1) test and implement protocols for long-term, multi-taxa monitoring of forests; 2) establish biodiversity assessment and monitoring projects to further regional conservation needs; 3) provide data management and analytical procedures that allow rapid assessment and dissemination of information, and 4) coordinate the interactive biodiversity monitoring network to facilitate information exchange and dissemination, and the formation of data quality standards.

In the late 1980s, the Southern Appalachian Biosphere Reserve was designated, along with establishment of the associated Southern Appalachian Man and the Biosphere (SAMAB) program and cooperative agreement codifying the collaborative intentions of several agencies (ultimately 14 federal and state agencies) (Van Sickle and Turner, 2001). SAMAB's vision is to foster a harmonious relationship between people and the Southern Appalachian environment. Over the course of its existence, SAMAB has garnered exemplary achievements in promoting environmental health and stewardship in natural and cultural ecosystems in the Southern Appalachian Mountains, and has demonstrated the remarkable potential that biosphere reserves have to bring people and resources together to achieve shared sustainability goals and objectives. Southern Appalachian Biosphere Reserve consists of six core areas including one national park (Great Smoky Mountains), two research areas, a state park, and two nature preserves. One of its greatest single contributions to date has been to lead the acclaimed Southern Appalachian Assessment, which provided a detailed accounting of the status of the Appalachian ecosystems and identified key areas of concern and opportunity to meet sustainability objectives throughout the ecoregion (SAMAB, 1996).

In the early 1990s, the Biosphere Reserve Inventory and Monitoring database was created to house and disseminate information on the biodiversity of biosphere reserves around the world. The utility and scope of the database eclipsed its original goal during the first decade of its existence to become a world-class resource as the Biological Inventory of the World's Protected Areas. The database is currently managed by the Information Center for the Environment's (ICE) at the University of California at Davis, and is readily accessed through the world wide web from any where in the world *(www.ice. ucdavis.edu).*

Throughout the 1990s, research directorates administered research that addressed one or two priority topics areas with respect to their ecosystem types, and together they completed seven major research projects addressing pressing sustainability challenges such as ecosystem management approaches to achieve ecological sustainability south Florida wetland ecosystems, caribou population and management dynamics in the arctic, ecological and socioeconomic impacts of management strategies for marine protected areas in three different oceanic areas, and the development of a land use change analysis system for modeling landscape-scale change at the ecoregional scale. In addition to large research projects, research directorates also distributed small grants to support student research projects on biosphere reserves.

The new millennium brought about many changes in the US MAB program. In 2000, administration of the US MAB program was transferred from the State Department to the US Forest Service, and the US MAB Secretariat and chair of the National Committee was conveyed to the Associate Deputy Chief for Research and Development in the US Forest Service. The Forest Service has been engaged in updating the US MAB program since 2000. In 2003, the United States rejoined UNESCO after a 19 year hiatus. Rejoining UNESCO brought with it renewed interest and optimism about the potential of the US MAB program to continue to make significant contribution to sustainability challenges in the 21st century.

The first US Biosphere Reserve Association was also formed in 2003 *(www.samab.org/about/usbra/usbra.html).* The US Biosphere Reserve Association is a non-profit organization dedicated to three primary aims: 1) provide leadership and support for the biosphere reserves; 2) convey factual information about the purposes and activities of biosphere reserves; and 3) develop cooperation among biosphere reserves in North America.

The US Biosphere Reserve Network

UNESCO regards the world network of biosphere reserves as MAB's most visible asset and operation tool in the 21st century, and a key mechanism by which priority work on sustainability issues will be accomplished. Similarly in the US, the biosphere reserve network is viewed as a collection of landscape for learning, which uniquely positions MAB to provide leadership and deliver substantive contributions toward sustainability in the US. At the global scale, external recognition for the unique value of the world network of biosphere reserves has come in the form of the prestigious Concord Award in 2001 from the Prince of Asturias Foundation. The award acknowledges its 30 years of contributions to the conservation of unique natural areas and associated species that are the heritage of mankind, and to opening new horizons of knowledge about how to protect and preserve ecological and cultural treasures *(www.fps.es/ing/premios/galardones/galardonconcordia2001.html).*

The US biosphere reserve network consists of 47 biosphere reserves (Figure. 2); about 60 percent were established in 1976 and the remaining ones were established over the following 15 years. Thirty-nine of the reserves are located across the contiguous US, with an additional four in Alaska, one in Hawaii, two in Puerto Rico, and one in the Virgin Islands. The 47 reserves span a diversity of ecosystem types, including terrestrial and marine ecosystems. The US biosphere reserve network is unique relative to networks in most other countries in a number of features. First, unlike most of the 393 reserves in other countries, almost all of the reserves in the US were established in areas that were already designated for conservation or research purposes (Table 1). Specifically, the majority (> 60 percent) of the biosphere reserves in the US are associated National Parks. USDA Experimental Forests are the next most prevalent association (17 percent), with the remainder located on a smattering of different lands, such as state forests,

Figure 2: US MAB Biosphere Reserve Network (from UNSECO)

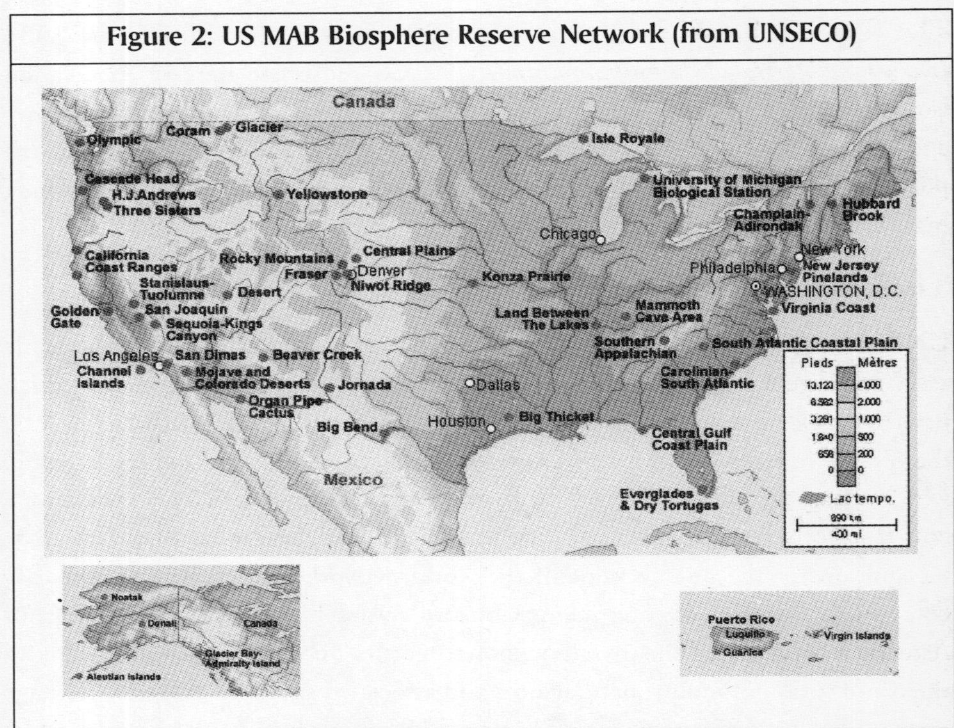

Table 1: Biosphere Reserves in the US MAB Network

Biosphere Reserve Name	Administrative Offices	Number of Units
Aleutian Islands	Adak, AK	1
Beaver Creek	Flagstaff, AZ	1
Big Bend	Big Bend, TX	1
Big Thicket	Beaumont	1
California Coast Ranges	Northern coast, CA	8
Carolinian-South Atlantic	Coast, SC	11
Cascade Head Biosphere	Corvalis, OR	1
Central Coast Biosphere	Stanford, San Francisco, Stinson Beach, Corte Madera, Burlingame, Glen Ellen, Bodega Bay, Neward, Novato, and Point Reyes, CA	14
Central Gulf Coast Plain	Eastpoint, FL	1
Central Plains	Ft. Collins and Nunn, CO	1

Contd...

Contd...		
Champlain-Adirondak	Ray Brook, NY and Waterbury and Rutland, VT	3
Channel Islands	Santa Barbara and Ventura, CA	2
Coram	Missoula, MT	1
Denali	Denali, AK	1
Desert Provo,	UT	1
Everglades & Dry Tortugas	Homestead, FL	1
Fraser	Ft. Collins, CO	1
Glacier Bay-Admiralty Island	Juneau and Gustavus, AK	2
Glacier	West Glacier, MT	1
Guanica Guanica,	Puerto Rico	1
H.J. Andrews	Corvalis, OR	1
Hawaiian Islands	Hawaii and Maui, HI	2
Hubbard Brook	Campton, NH	1
Isle Royale	Houghton, MI	1
Jornada Las	Cruces, NM	1
Konza Prairie	Manhatten, KS	1
Land Between the Lakes	Golden Pond, KY	1
Luquillo Rio Piedras,	Puerto Rico	1
Mammath Cave Area	Bowling Green, KY	1
Mojave and Colorado Deserts	Palm Desert, Borrego Springs, Death Valley, San Bernardino, and Twentynine Palms, CA	5
New Jersey Pinelands	New Lisbon, NJ	1
Niwot Ridge	Nederland, CO	1
Noatak	Fairbanks and Kotzebue, AK	2
Olympic	Port Angeles, WA	1
Organ Pipe Cactus	Flagstaff and Ajo, AZ	1
Rocky Mountain	Estes Park, CO	1
San Dimas R	iverside, CA	1
San Joaquin	Fresno, CA	1
Sequioa-Kings Canyon	Three Rivers, CA	1
South Atlantic Coastal Plain	Hopkins and Georgetown, SC	1
Southern Appalachian	Knoxville, TN	6
Stanislaus-Tuolumne	Sonora, CA	1
Three Sisters	McKenzie Bridge, OR	1
Univ. of Michigan Biological Station	Ann Arbor, MI	1
Virgin Islands	St. John, Virgin Islands	1
Virginia Coast	Nassawadox, VA	1
Yellowstone	Jackson Hole, WY	1

US Fish and Wildlife National Wildlife Refuges, Bureau of Land Management lands, University properties, and Nature Conservancy properties. Thus, biosphere reserve designations did not increase the amount of land set aside for conservation in the US, but rather served to strengthen and broaden the emphasis of unit management to encompass research, monitoring, development, education, and demonstration of sustainability practices. Second, a large proportion (>30 percent) of reserves consists of multiple core areas (in other words, individual administrative units). Reserves with multiple units serve an important function in the US MAB network, because they present a complex administrative scenario where multiple and potentially conflicting mandates of various units need to be considered and management options are negotiated to best meet the shared objectives and individual needs of each unit. Thus, multiple unit reserves have some of the same management challenges as buffer and transition areas. Finally, most US biosphere reserves do not have buffer or transition areas designated. This is largely a function of the reserves being based on previously designated areas that only consisted of core reserve areas with a single set of objectives. A few of the reserves (for example, Mammoth Cave Area Biosphere Reserve) have proposed zones of cooperation, which are intended to serve a similar function as buffer and transition areas.

In some respects, reserves without designated buffer and transition areas serve as ecological and social experiments in the value and necessity of buffer and transition areas in meeting the objective of maintaining biological diversity of core areas. The evolution of boundary designations and land management in and around Yellowstone National Park is a prime example. Yellowstone National Park was the first national park in the US, established in 1872 (USDI, 2004), and it was designated as a biosphere reserve in 1976 with only the core area of the park designated (no buffer or transition areas). The park itself extends over a 9,00,000 ha area. The concept of a Greater Yellowstone Ecosystem came into being in the mid 1980s, driven by the recognition that many of the values for which the park was established were dependent upon the management of lands outside the park. As such, the Greater Yellowstone Ecosystem was established to maintain the important ecological and social linkages across the headwaters of three major river systems – Yellowstone, Snake, and Green rivers – and thereby maintain ecosystem integrity, including populations of wide ranging species that require large landscapes to persist, such as grizzly bear, wolverine, lynx, and gray wolf.

The Greater Yellowstone Ecosystem currently occupies a 7 million ha area centered on Yellowstone National Park, including two National Parks, portions of six national forests, three national wildlife refuges, BLM lands, state lands, and private and tribal lands (Clark and Minta, 1994; Shullery, 1995). A future review of the Yellowstone Biosphere Reserve is likely to conclude that the boundaries of the Greater Yellowstone Ecosystem function as a zone of cooperation and should be designated as such within the MAB program.

UNESCO states that biosphere reserves are to be reviewed every 10 years to evaluate their effectiveness and needs. Toward this end, a survey of biosphere reserves in the US was conducted in 1995, and repeated in 2003, to determine their functional strengths and needs (Gilbert, 2004). The survey asked each reserve to provide their perspective on the state and future of US biosphere reserves. Five primary questions were asked of managers: 1) what are the management benefits to your biosphere reserve, 2) who is participating, 3) what resources are needed, 4) who identifies with the biosphere reserve concept, and 5) who is concerned about or opposed to the biosphere reserve. A number of substantive conclusions came from the most recent survey. Biosphere reserve status conferred management benefits for most reserves, and benefits were greatest in the areas of promoting an ethic of sustainability, research, public consultation and participation, environmental awareness, ecosystem management, and improved cultural resource protection. Academic research institutions were the most prevalent participants in the biosphere reserve activities, followed by federal agencies. Few units experienced expressions of concern by local communities and organizations. Perhaps most importantly but not surprising is that funding was the most limiting factor, followed by limited local involvement and infrequent communication among biosphere reserve managers. Without funding from the MAB program, the mission of units tends to shrink back to the core mission of the unit's original designation, and most frequently that means that programs directed toward sustainable development atrophy. Finally, responses were quite similar in 1995, with some changes in the focus of programs. Since 1995, areas of emphasis for biosphere reserves have shifted from conservation and ecosystem management to research and education.

The ecological status of US biosphere reserves has never been evaluated as a network. Given that most biosphere reserves are co-located with national parks, we consulted the results of the National Parks Conservation Association's (NPCA) annual evaluation of national parks *(www.npca.org)* as a small window into their status. Every year the NPCA identified the 10 most endangered National Parks out of the 387 units designated in the National Park system. In 2003, 5 of the 10 most endangered National Parks are also biosphere reserves (Big Thicket National Preserve, Everglades National Park, Great Smokey Mountains National Park, Organ pipe cactus National Monument, and Yellowstone National Park). Predictably, biosphere reserves co-located with National Parks will suffer from the same set of stressors that affect the parks, namely air and water pollution and recreation impacts. A more thorough evaluation of the status of the ecological status of biosphere reserves would be a valuable tool to guide future investments in biosphere reserve management and considerations for additions to the network.

Future Directions

The US MAB program is in a period of reflection and revitalization as it contemplates its future commitment and contributions to biological diversity conservation and sustainable development. Much has transpired over the last three decades, including advancement of our understanding of ecology, ecosystem dynamics, and sustainability thresholds and threats. Similarly, many national and international programs have been created with the intent of achieving a variety of biodiversity conservation and sustainability objectives (for example, the Long-term Ecological Research Network and the National Center for Ecological Analysis and Synthesis). At the same time, it is clear that sustainability challenges continue to mount. At the Johannesburg World Summit on Sustainable Development in 2002, it was determined through exchange and discussion by thousands of participants, including over 100 heads of state, that growing poverty and increasing environmental degradation are threatening sustainability at local to global scales. The Summit intended to focus the world's attention toward meeting the challenge of improving people's lives and conserving natural resources in the face of growing human populations and association demands for resources. Given the limited progress that has been made since the Rio de Janeiro Summit in 1992, there is now an urgency to take substantive action, and the US MAB program could play a pivotal role.

It is incumbent upon the US MAB program to take a multitude of factors into consideration in charting its course for the future. Certainly, the draft recommendations from the Johannesburg World Summit will outline keystone steps in the pursuit of sustainability. Similarly, the 2002 Convention on Biological Diversity generated broad objectives for 2010 that mirror those of the 2002 World Summit on Sustainable Development – poverty reduction and slowing the current rate of biodiversity loss at the global, national, and regional levels (*www.biodiv.org*). Specific approaches and actions have been generated by the MAB program through two seminal conferences. MAB program representatives from over 40 countries convened in Seville, Spain in 1995 to examine past experience in implementing the innovative concept of biosphere reserves, and to look to the future to identify what emphasis should now be given to their three functions of conservation, development and logistical support. The result was Seville Strategy, which contains 11 objectives and associated recommendations for improving the national and bioregional effectiveness of the MAB program (Table 2) (UNESCO, 1996). A subsequent meeting in 2000 resulted in another document that provides further refinement to the original strategy in the form of a checklist for priority actions referred to as Seville+5 (UNESCO, 2000). These recommendations are tailored specifically to the MAB program and provide a clear set of actions that will assist in directing and energizing the US Program.

Table 2: Four Goals and 11 Objectives Articulated in the Seville Strategy (recommendations for action are available)

Goal I: Use of biosphere reserves to conserve natural and cultural diversity

- Improve the coverage of natural and cultural biodiversity by means of the World Network of Biosphere Reserves.
- Integrate biosphere reserves into conservation planning.

Goal II: Utilize biosphere reserves as models of land management and of approaches to sustainable development

- Secure the support and involvement of local people.
- Ensure better harmonization and interaction among the different biosphere reserve zones.
- Integrate biosphere reserves into regional planning.

Goal III: Use biosphere reserves for research, monitoring, education, and training

- Improve knowledge of the interactions between humans and the biosphere.
- Improve monitoring activities.
- Improve education, public awareness, and involvement.
- Improve training for specialists and managers.

Goal IV: Implement the biosphere reserve concept

- Integrate the functions of biosphere reserves.
- Improve the strength of the World Biosphere Reserve Network.

In summary, retooling the US MAB program for success over the next few decades will undoubtedly updatie many of its elements. A refreshed set of objectives and operating principles based on a model of integrated human and natural ecosystems is needed. An increased emphasis on the interface between human-dominated and wildland ecosystems will guide revisions and additions to the reserve network (for example, urban biosphere reserves), as well as within-reserve activities. Enhanced networking capabilities among reserves within the US and with other countries is critical to speed the exchange of ideas and the application of innovations. Global climate change has been identified as a sustainability challenge that biosphere reserves can make a unique contribution, as evidence by the recent MAB Global Change Monitoring Initiative (UNESCO, 2004). The US biosphere reserve network contains a number of high elevation and island ecosystems, and has the opportunity to contribute to global change challenges. Finally, national direction and support is essential to promote and advance much needed education and capacity building activities at multiple scales. The US MAB program is poised at the event horizon of a new century of promise and challenges, and holds great potential to contribute substantially to the goals, objectives, and challenges in the pursuit of sustainability in the decades ahead.

Acknowledgement

Manley P N and D C Hayes 2005. Biodiversity conservation, sustainable development, and the US Man and the Biosphere Program: Past contributions and future directions. In C Aguirre-Bravo and others, eds. Monitoring science and technology symposium: Unifying knowledge for sustainability in the Western Hemisphere. 2004 September 20-24; Denver, CO.

Proceedings RMRS-P-37CD. Fort Collins, CO: US Department of Agriculture, Forest Service, Rocky Mountain Research Station. CD-ROM.

(P N Manley, Research Wildlife Biologist and Technical Assistant to US MAB Coordinator, US Forest Service, Pacific Southwest Research Station, Davis, CA and D C Hayes is the National Program Leader for Watershed Research and US MAB Coordinator, USDA Forest Service, Washington DC.)

References

Anonymous 1995, Biosphere reserves in action: Case studies of the American experience. Unpublished document. US MAB Secretariat, US Department of Agriculture, Forest Service, Washington, D.C. *(www.euromab.org/publications/ case.html).*

Clark T W; S C Minta 1994, Greater Yellowstone's future. Homestead Publishing, Moose, WY. 160 p.

Diamond J 1999, Guns, germs, and steel: the fates of human societies. W W Norton & Co., New York, NY. 480 p.

Gilbert T 2004, United States Biosphere Reserve Survey 2003. Unpublished report. United States Biosphere Reserves Association, Knoxville, Tennessee. *(www.samab. org/about/usbra usbra.html).*

Southern Appalachian Man and the Biosphere (SAMAB). 1996. The Southern Appalachian Assessment Summary Report. Report 1 of 5. US Department of Agriculture, Forest Service, Southern Region, Atlanta, GA. 128 p.

Sayer J; B Campbell 2003, The science of sustainable development: local livelihoods and the global environment. Cambridge University Press, Cambridge, United Kingdom. 288 p.

Schullery P 1995, The Greater Yellowstone Ecosystem. In LaRoe, E T; G S Farris; C E Puckett; P D Doran; M J Mac, eds. Our living resources: a report to the nation on the distribution, abundance, and health of US plants, animals, and ecosystems. US Department of the Interior, National Biological Service Washington, D.C. 312 to 314.

UNESCO 1995, Statutory framework of the world network of biosphere reserves. UNESCO, Division of Ecological Sciences, Paris, France. 6 p. *(www.unesco.org/mab/docs/ statframe.html).*

UNESCO 1996, The Seville Strategy for Biosphere Reserves. United Nations Educational, Scientific and Cultural Organization, Paris, France. 13 p. *(www.unesco.org/mab/ publications/ document.html).*

UNESCO 2000, Seville+5 Recommendations: Checklist for Action. United Nations Educational, Scientific and Cultural Organization, Paris, France. 11 p. *(www.unesco. org/mab/ publications/document.html).*

UNESCO 2004, Global change in mountain biosphere reserves proceedings. International workshop on global change in mountain biosphere reserves. Entlebuch biosphere reserve, Switzerland. November 10-13, 2003. 20 p.

UNESCO 2003, Biosphere Reserves list. UNESCO MAB Secretariat, Division of Ecological Sciences, Paris, France. 19 p. *(www.unesco.org/mab/brlist.html).*

US Census Bureau. 2004. Global population at a glance: 2002 and beyond. International brief WP/02-1, March 2004. 4 p. *(www.census.gov/prod/2004pubs/wp02-1.pdf; also see www.census.gov/popest/estimates.php).*

USDI 2004, Yellowstone Resources and Issues 2004: a compendium of information about Yellowstone National Park. Unpublished document. National Park Service, Washington, D.C. 88p. *(www.nps.gov/yell/publications/pdfs/handbook/ index.html).*

Van Sikle C;R S Turner 2001, The Southern Appalachian Man and the Biosphere program – a model for management need-based research. Unpublished document. Presentation to the Southern Forest Science Conference, Atlanta, GA, November 2001. SAMAB, Knoxville, TN. *(www.samab.org).*

UNESCO's MAB Program: An Overview

UNESCO's Man and the Biosphere Program has initiated programes and activities focusing on the diversity and the resources provided by nature, humans' impacts on biodiversity, as well as how biodiversity affects human activities. These initiatives are intended to contribute to the fulfillment of a global biodiversity agenda.

The key aspects considered under UNESCO's MAB Program are the following:

1. **Biodiversity Science and Policy:** Biodiversity is a complex area, and there are many scientific aspects of it that require further elucidation. Scientific information on biodiversity is important to help to build the basis on which to take informed policy decisions.
2. **Biological and Cultural Diversity:** Natural systems cannot be understood, conserved and managed without the recognition and respect of the human cultures that shape them. Together, biological and cultural diversity hold the key to ensuring resilience in both ecological and social systems and understanding the links between nature and culture is crucial for its safeguard. UNESCO's involvement in sacred natural sites and cultural landscapes as areas of biodiversity conservation is an obvious outcome of the shared work it carries out in the natural sciences and culture sectors. Another good example of the strong linkages between cultural heritage and biological diversity conservation can bee seen in the project "Quranic Botanic Gardens Network" in the Arabian Peninsula developed by the UNESCO Doha Office.
3. **Biodiversity Education:** UNESCO works with a number of constituencies in promoting education and outreach on biodiversity, combating desertification, climate change, ecosystem services and other issues that are key to achieving sustainable development. UNESCO is also the United Nations designated leader agency of the Decade of Education for Sustainable Development (2005-2014).
4. **Saving Great Apes:** Gorillas, chimpanzees, bonobos and orangutans are on the very edge of extinction. Yet their survival is directly depending on our capacity to counteract habitat fragmentation – one of the main causes of biodiversity loss. UNESCO has united with key international partners to address the crisis.

The MAB Program is supported by regional or sub-regional networks, and the UNESCO regional offices also play a vital role in the everyday implementaion of its activities activities. Here you will find links to MAB's regional networks.

Sources:1. http://www.unesco.org/mab/biodiv.shtml
2. http://www.unesco.org/mab/networks.shtml

16

Climate Change and Biodiversity in Europe

Hannah Reid

Climate change is already affecting European biodiversity, as demonstrated by changes in species' ranges and ecosystem boundaries, shifts in reproductive cycles and growing seasons, and changes to the complex ways in which species interact (predation, pollination, competition and disease). These effects vary between regions and ecosystems. Strategies adopted to mitigate or adapt to climate change also impact biodiversity. And land use changes with subsequent changes to biodiversity can also alter levels of greenhouse gas emissions thus affecting the global climate. This article describes the many linkages between climate and biodiversity. It stresses the need for more integrated policy responses, at international, regional and national levels. Nationally, activities that meet biodiversity and climate change objectives need promoting and mainstreaming into various sectoral areas of policy making. Reducing energy consumption, increasing energy efficiency and promoting renewable energy technologies are priorities. Activities which meet both climate and biodiversity objectives also need exploring. These include adopting new landscape

Source: Conservation and Society, March 2006. © Hannah Reid, 2006. Reprinted with permission.

management approaches, and ensuring that carbon sequestration projects and renewable energy projects incorporate biodiversity considerations.

Introduction

The earth's climate has gone through many periods of significant warming or cooling throughout its history. The fossil record demonstrates that these changes have had important ecological impacts, and have at times coincided with a number of mass extinctions (IPCC, 2001). Current global warming, however, is unusual. It is fuelled mostly by human activities; especially greenhouse gas emissions from burning fossil fuels. Agriculture and land use changes, and other industrial processes that release greenhouse gases also contribute to climate change (IPCC, 2001). Carbon dioxide concentrations in the lower atmosphere have increased from pre-industrial concentrations of 280 parts per million to 375 parts per million concentrations in 2003. This is the highest level in the last 50,000 years (IPCC 2001). These changes in carbon dioxide and other greenhouse gas concentrations have led to changes in climate.

The climate is warming in most parts of the world. The global average temperature has increased by about 0.7°C and the European average by about 0.95°C in the last 100 years (EEA, 2004). It is estimated that temperature will increase by 1.4-5.8°C globally and 2.0-6.3°C in Europe by 2100 (EEA, 2004). Precipitation patterns are more varied. In the last 100 years northern Europe has become 10-40% wetter and southern Europe up to 20% drier. These changes are projected to continue (EEA, 2004). The Intergovernmental Panel on Climate Change (IPCC, 2001) states that 'the risk of water shortage is projected to increase particularly in southern Europe.... Climate change is likely to widen water resource differences between northern and southern Europe.' Eight out of nine glaciated regions in Europe have shown significant recent retreat. From 1850 to 1980 glaciers in the European Alps lost about one third of their area (EEA, 2004). In the past 100 years, European sea levels have risen by 0.1 to 0.2 m (IPCC, 2001). Currently the sea level around European coasts is rising at a rate of 0.8 mm/year (in Brest and Newlyn) to 3.0 mm/year (in Narvick) (Liebsch *et al.*, 2002).

In addition, extreme weather events such as droughts, heatwaves and floods have increased in Europe, while cold extremes have decreased (Klein Tank, 2004).

The extent and rate of current climate change exceeds all natural variation in the last 1000 years and possibly further back in history (Houghton *et al.*, 2001). The speed of change makes it hard for humans, other species and ecosystems to adapt. The modern landscape also provides little flexibility for ecosystems to adjust to rapid environmental change. In contrast with historical migration responses, species today must move through a landscape that is increasingly impassable due to the widespread loss and fragmentation of habitats (Kappelle *et al.*, 1999; Walther *et al.*, 2002). Indeed, the IPCC (2001) states that in Europe, 'adaptation potential for natural systems is generally low' in part because 'Europe is predominantly a region of fragmented natural or semi-natural habitats in a highly urbanised, agricultural landscape.'

The debate about climate change has now reached a stage where most scientists accept that, whatever happens to future greenhouse gas emissions, we are now locked into a future characterised by significant human-induced changes to our climate. There are two types of response to these changes: the first is to try and reduce the extent to which our climate is altered. This is known as climate change mitigation. The second is to learn to live with the inevitable changes. This is known as adaptation to climate change. Biodiversity is inextricably linked to climate. Changes in climate affect biodiversity and changes to natural ecosystems affect climate. This article considers the linkages between climate change (mitigation and adaptation) and biodiversity. It then discusses the need for integrated policy responses, and ends with some practical ways forward to provide both biodiversity and climate change benefits on the ground.

Direct Impacts of Climate Change on Biodiversity

This section discusses the most important direct impacts that climate change could have on biodiversity.

Species Ranges

Climate change is likely to have a number of impacts on biodiversity—from ecosystem to species level. The most obvious impact is the effect that temperature and precipitation changes have on species' ranges and ecosystem boundaries. Any particular ecosystem consists of an assemblage of species, some of which will be near the edge of their ranges and others will not. Those at the edge of their ranges may need to move due to climate change (Leemans and van Vliet, 2004). Species that are highly mobile or opportunistic are likely to benefit at the expense of those that are not (Hughes, 2000; Secretariat of the CBD, 2003). The invasion of alien species may be facilitated in some parts of Europe. Root *et al.*, (2003) found that out of 1,700 species reviewed globally, 81% of observed shifts in range were in the direction expected by climate change. In Europe, Bakkenes *et al.*, (2002) predict that under existing climate change forecasts only 32% of plant species existing in an average 'grid cell' in 1990 would still be present in 2050.

In terrestrial Europe, trees and shrubs have replaced many Arctic and tundra communities (Molau and Alatalo, 1998). The treeline and the level at which alpine plants are found in Europe is moving towards higher altitudes (Walther *et al.*, 2002). In Sweden, the treeline is predicted to rise by 233 to 677 m depending on the climate scenario and location (Moen *et al.*, 2004). European tree species may be replaced by new species that are better adapted to higher temperatures and drought stress. Plants or trees that need low temperatures in winter to trigger bud bursting in spring may be adversely affected (Chuine and Beaubien, 2001). In the Netherlands, thermophilic plant species have become 60% more common compared with 30 years ago, and cold tolerant species have declined (Tamis *et al.*, 2001).

European insects, mammals and diseases also demonstrate range shifts (Leemans and van Vliet, 2004). For example, out of 35 butterfly species in Europe, 63% have shifted their ranges northwards by 35-240 km over the last century, whereas only 3% have shifted south (Parmesan *et al.* 1999). In Finland, macrolepidopteran species richness is expected to increase as southern species shift their ranges northwards (Virtanen and Neuvonen, 1999). Conversely, the distribution of northern species, comprising 11% of Finnish species, may shrink (Virtanen and Neuvonen, 1999). Thomas and Lennon (1999) report that the ranges of many British bird species have moved an average of 18.9 km northwards in the last 20 years.

In European seas, temperate species have migrated about 250 km northward per decade (Parmesan and Yohe, 2003), whereas sub-Arctic and Arctic species have declined in number (Beaugrand *et al.* 2002). More temperate species are now found in the North Sea (EEA, 2004). European seas have shown surface temperature increases, especially in isolated basins like the Baltic Sea and the North Sea (EEA, 2004). This has increased phytoplankton biomass, and moved the ranges of zooplankton species by up to 1000 km in the last few decades (EEA, 2004) Clark *et al.* (2003) predict that the North Sea cod population will increasingly decline due to higher sea temperatures.

Shifts in ecosystem boundaries could mean that protected areas, such as the Swiss National Park, no longer contain the species and habitats they were established to protect. The Pasterze Glacier has also retreated several hundred metres since the 1970s, thus affecting the Hohe Tauern National Park in Austria (Dudley, 2003). Under existing static conservation paradigms, little emphasis is placed on changing patterns of biodiversity. And few protected area systems have been formulated with reference to climate change, even in countries where effects will probably be large (Hannah *et al.*, 2002). The World Wide Fund for *Nature* (WWF) argues that protected areas offer limited defence against problems posed by rapid environmental change, and that protected areas themselves will need to adapt to meet the challenges posed by global warming (Dudley, 2003).

Flooding and sea level rise will affect species' ranges and ecosystem boundaries as well as threatening wetlands and coastal ecosystems (Dudley 2003). Man-made coastal defences across the UK are eroding faster than ever and destroying important inter-tidal habitats for birds (RSPB, 2004a). The IPCC (2001) states that in European 'coastal areas, the risk of flooding, erosion, and wetland loss will increase substantially... southern Europe appears to be more vulnerable to these changes, although the North Sea coast already has high exposure to flooding.'

Changes in Phenology

Climate change is also causing shifts in the reproductive cycles and growing seasons of certain species. This can alter the frequency of pest and disease outbreaks. Research by Parmesan and Yohe (2003) on the timing of spring events, such as egg laying by birds or flowering by plants, showed that in 61 studies, the timing had shifted earlier by an average of 5.1 days per decade over the last half century. Likewise, Root *et al.* (2003) found that out of 1,700 species reviewed, 87% of observed shifts in phenology were in the direction expected by climate change. In Europe, phenological data shows an increase in the length of the growing season by about 10 days from 1962 to 1995 (Menzel and Fabian, 1999). Leemans and van Vliet (2004) note phenological shifts in certain mammal, reptile, fish and bird species. For example, bird life cycle events such as migration and egg laying appear to be occurring earlier (Crick *et al.*, 1997). Increasingly warm winters associated with the North Atlantic Oscillation influence the development and fecundity of red deer (Cervus elaphus) in Norway (Post *et al.*, 1997). And in the UK, amphibians are spawning nine to ten days earlier than they were 17 years ago (Beebee, 2002).

Changes in Species Interactions

The impact that climate change will have on many of the more complex interactions (predation, competition, pollination and disease) that constitute functioning ecosystems remains largely unknown (Walther *et al.*, 2002; Leemans and van Vliet, 2004). However, some studies provide indications of what might be expected. For example, the indirect effects of climate change can be seen in a food chain in the North Sea where warmer water is driving cold water plankton further north. This reduces the survival of young cod and sandeels. A sandeel shortage may in turn have contributed to the large breeding failure seen in seabirds on the North Sea coast of Britain in 2004 (Lanchbery, 2004; RSPB, 2004a).

Changes in competitive ability may also emerge. For example, early leafing trees will get a two-week head start on their competitors in terms of growth, and thus occupy an increasing proportion of woodland (Dudley, 2001). Numbers of long distance migrating birds in the Lake Constance region (a Ramsar site at the border of Austria, Germany and Switzerland) have declined, probably due to warmer winters increasing the competitive advantage of resident bird species (Lemoine and Böhning-Gaese, 2003).

Mismatches in timing between interdependent species may occur, especially when changes in some species are cued by day length, and others by temperature (Hughes, 2000). For example, in one wood in Oxfordshire, UK, there is evidence that blue tit hatching no longer coincides with peak caterpillar numbers (Dudley, 2001).

Disease and predation may also increase. For example, numbers of native pests such as the green spruce aphid may increase due to milder winters, and non-native pests such as the pinewood nematode, gypsy moth and Asian longhorn beetle are now found in southern British woods (Broadmeadow, 2000; Dudley, 2001).

Extinction Rates

Climate change may lead to a sharp increase in rates of extinction. Thomas *et al.*, (2004) studied five regions of the world, and predicted that if the present rate of climate change continues, 24% of species in these regions will be on their way to extinction by 2050. This study indicated that for many species, climate change poses a greater threat to their survival than the destruction of their natural habitat. Evidence directly linking climate change and species extinction is difficult to procure, but at least one species: the golden toad of Costa Rica, may have become extinct due to climate change (Pounds *et al.*, 1999).

Effects of Extreme Climate Events

Climate change related extreme events such as disease, drought, fire or an El Niño are likely to increase, and can seriously affect biodiversity (Hannah *et al.*, 2002). These extreme events may affect organisms, populations and ecosystems more than gradual global or regional changes in averages (Leemans and van Vliet, 2004; Walther *et al.* 2002). For example, Spain lost more than 485,622 hectares of forest to wildfires in 1994 and Italy lost 149,734 hectares in 1998 (Pinol *et al.*, 1998). The 1976 drought in the UK severely affected tree health, and vulnerable species like beech took years to recover (Dudley, 2001). About 15 million trees were blown down in the UK during the October 1987 storm (IPCC, 2001). Broadmeadow (2000) predicts that higher mean wind speed and an increase in the occurrence of storms will make British woodlands more vulnerable to wind damage.

Variations between Regions, Species and Ecosystems

The impacts of climate change on biodiversity will vary between regions. The most rapid changes in climate are expected in the far north and south of the planet, and in mountainous regions (Pauli *et al.*, 2001). Unfortunately, these regions are also home to many species which have no alternative habitats to which they can migrate in order to survive, and species which cannot easily compete with new immigrating species (Pauli *et al.*, 2001). In Europe, the greatest effects of climate change are projected for Arctic Regions, the moisture-limited ecosystems of eastern Europe, and the Mediterranean region (Bakkenes *et al.*, 2002). The IPCC (2001) states that 'the Arctic is likely to respond rapidly and more severely than any other area on earth, with consequent effects on sea ice, permafrost, and hydrology'. It adds that 'polar warming probably should increase biological production, but different species compositions are likely on land and in the sea, with a tendency for poleward shifts in major biomes and associated animals'. Theurillat and Guisan (2001) state that high mountain systems such as the Alps will also be particularly vulnerable.

Some species and ecosystems are also more vulnerable than others, particularly small populations or those restricted to small areas. For example, endangered birds in the UK, especially those living in fragile mountain habitats, such as the dotterel (Charadrius morinellus), ptarmigan and snow bunting, may have nowhere to move to and be lost from the UK forever (RSPB, 2004a). In mountainous regions of Europe, endemic tree species have been replaced by other species (such as spruce and pine), which have migrated upwards due to a number of factors including climate change (Pauli *et al.*, 2001). The IPCC, (2001) states that in Europe, 'Tundra areas have practically no adaptive options available'.

Indirect Impacts of Climate Change on Biodiversity

It is not just climate change itself that can impact biodiversity. In some cases, the strategies that are adopted to mitigate or adapt to climate change can affect biodiversity. This section discusses the most important indirect impacts that climate change could have on biodiversity.

Classic top-down approaches to climate change often equate to large infrastructure construction projects. Projects designed to support adaptation to climate change are often associated with physical protection. For example, large sea walls may be built to protect against storm surges and floods. These are particularly important in Asia, and the Netherlands continues to invest millions in improving its sea defences against sea level rise and storms (Mortished, 2005). Such projects often negatively impact biodiversity. Alternative adaptation options, such as strategic placement of artificial wetlands (or protection of existing mangroves or coral reefs outside Europe) can benefit biodiversity (Secretariat of the CBD, 2003). Other adaptation projects involve using pesticides and herbicides to control pests and diseases that might increase due to climate change. Use of such chemicals may damage existing plant and animal communities (Secretariat of the CBD, 2003).

Projects designed to reduce global greenhouse gas emissions and thus mitigate climate change are often associated with large renewable energy schemes. These often have poor outcomes for biodiversity. For example, large hydropower schemes can cause loss of terrestrial and aquatic biodiversity and inhibit fish migration (Fearnside, 2001; Fu *et al.*, 2003). Locating dams near river estuaries is particularly damaging, as it disrupts the whole watershed. Dams can also be net emitters of greenhouse gases if submerged soils and vegetation decay and release carbon dioxide and methane (Secretariat of the CBD, 2003; World Commission on Dams, 2000). Europe accounts for 70% of the total installed wind energy capacity in the world (Secretariat of the CBD, 2003). Wind farms pose three main problems for birds: disturbance, habitat loss/damage, and collision. For example, wind turbines in Tarifa and Navara in Spain have killed many raptors. The design, location and management of wind farms can limit these problems (RSPB, 2004b). The effect that offshore wind farms will have on sea mammals, fish and marine aquatic communities is largely unknown (Secretariat of the CBD, 2003).

Afforestation and reforestation activities designed to sequester carbon and therefore mitigate climate change can restore watershed functions, establish biological corridors and provide considerable biodiversity benefits if a variety of different aged native tree species are planted. Monocultures, however, not only reduce biodiversity, but also increase the chances of pest attacks, thus challenging the permanence of carbon stocks. The location of afforestation and reforestation projects is also important. Replacing native grasslands, wetlands, shrublands or heathlands may lead to dramatic biodiversity losses, and also lower the relative increase in carbon sequestered compared to implementing such projects on degraded land (IUCN, 2004; Reid, 2003; Secretariat of the CBD, 2003).

How Ecosystems and Biodiversity Affect Climate

Just as climate change affects biodiversity, so changes in biodiversity can affect the global climate. This section documents some of the most important ways in which biodiversity affects the climate.

Land use changes and subsequent changes to biodiversity can both increase and decrease greenhouse gas emissions. Countries like Ireland, the Netherlands and Denmark are significant carbon sources, but countries like Slovenia and Slovakia are significant carbon sinks (Janssens *et al.*, 2003).

Forests are a major carbon store. Carbon dioxide is released whenever there are forest fires, or when forests are cut down. Globally, deforestation, mainly in tropical regions, is thought to be responsible for annual emissions of 1.1 to 1.7 billion tonnes of carbon per year, or approximately one fifth of human carbon dioxide emissions (Brown *et al.*, 1996). European forests were almost completely cleared during the agricultural expansion of the 16th to 18th centuries. This released considerable quantities of carbon from the vegetation and soil. European forests currently store on average 70-160 g of carbon per square metre per year, 70% of which is in trees, and 30% in soils. This makes them an important carbon sink. During the 1990s the European terrestrial biosphere stored 7-12% of the annual anthropogenic carbon dioxide emissions (Janssens *et al.*, 2003). Nabuurs *et al.*, (2002) predict that this is likely to continue, as climate change will increase the quantity of stemwood in European forests with an additional 0.9 m^3 per hectare per year in 2030.

Peatlands provide many environmental services, such as improving water quality. Many are important biodiversity reservoirs or stopover points for migratory species. Peatlands also hold roughly one-third of the soil carbon worldwide, and greenhouse gases are released every time they are burned, drained, converted to agriculture or degraded. Peatland forest fires in Indonesia during 1997 released an amount of carbon dioxide equivalent to 40% of the world's average yearly carbon emissions from fossil fuels (Page *et al.*, 2002). In Europe, peat extraction for horticulture, agriculture and energy is reducing stored carbon amounts. In Russia, permafrost peatlands appear to be melting and drying out and fire frequency is increasing (Janssens *et al.*, 2003). Melting permafrost and the consequent release of greenhouse gases from northern wetlands can enhance climate change (Secretariat of the CBD, 2003). Overall, European peatlands are a net source of carbon to the atmosphere of 70 Tg of carbon per year, equivalent to about 20% of the carbon sequestered by the European forest sector (Janssens *et al.*, 2003).

Currently, some 60% of anthropogenic global greenhouse gas emissions originate from the generation and use of energy. While the use of renewable energy sources, such as wood, instead of fossil fuels can help mitigate climate change, this can also have negative impacts. For example, using relatively undisturbed natural fuel sources (such as native forests as opposed to plantations) can lead to significant biodiversity losses. Some bio-energy plantations replace sites with high biodiversity, introduce alien species and use damaging agrochemicals (Secretariat of the CBD, 2003).

There are also many complex feedback mechanisms at work between biodiversity and climate change. For example, some species of ocean algae release dimethylsulfide into the atmosphere. Rising ocean temperatures due to global warming can lead to algal blooms, and resultant increases in dimethylsulfide release contribute to cloud formation. This in turn may help reduce temperatures, as less heat will be able to reach the earth's surface (Sciare *et al.*, 2000). Rising temperatures can stimulate the growth of phytoplankton in the sea, which can increase carbon dioxide uptake. Likewise higher atmospheric carbon dioxide concentrations and warmer temperatures can stimulate forest growth and hence carbon dioxide uptake in mid and northern European forests (Broadmeadow, 2000; EEA, 2004). In high latitudes, the replacement of shrub/tundra vegetation with trees can affect the radiation balance. Unlike snow, trees are more likely to absorb sunlight rather than reflect it. This in turn can enhance climate change (Secretariat of the CBD, 2003).

Integrated Policy Responses

The many ways in which climate and biodiversity interact suggest a need for more integrated policy responses. Synergies between the United Nations Framework Convention on Climate Change (UNFCCC) and the Convention on Biological Diversity (CBD) need to be explored. The Ad Hoc Technical Expert Group on Biological Diversity and Climate Change has identified some possible linkages (Secretariat of the CBD, 2003), and the convention secretariats have established a joint liaison group to help link initiatives relating to climate change and biodiversity. However, getting those responsible for implementing the two conventions to work together is difficult. The conventions have separate constituencies, administration arrangements, negotiators and guiding scientific bodies. Encouraging countries to set up a single institution to deal with obligations under all international environmental agreements could be one way forward (Reid, 2004; Reid *et al.*, 2004b).

The UNFCCC (in Article two) states that the ultimate objective of the convention is to ensure atmospheric greenhouse gas concentrations stabilise 'within a time-frame sufficient to allow ecosystems to adapt naturally to climate change'. Parties to the UNFCCC are guided by the principle that carbon sequestration activities should also contribute to biodiversity conservation and the sustainable use of natural resources. But common guidelines on what this timeframe is or how these principles might be realised have not been agreed. The Kyoto Protocol and Marrakech Accords (under the UNFCCC) do not explicitly exclude practices such as afforestation of native grasslands or wetlands (IUCN, 2004). The Clean Development Mechanism (CDM), established under the Kyoto Protocol, helps developed countries to meet emissions reduction targets, by allowing them to take credits from emissions reduction projects in poor nations. Projects are supposed to provide global benefits from carbon sequestration, but also sustainable development benefits. These benefits could actively incorporate biodiversity conservation, soil protection and other environmental concerns (IUCN, 2004; Huq and Reid, 2005). However, without a minimum set of common international standards, CDM and other climate change mitigation projects could flow to countries with minimal standards, thus adversely affecting biodiversity (Secretariat of the CBD, 2003). The UNFCCC also currently classifies harvested forest products as emissions as soon as they leave the forest site. This fails to recognise the value of carbon sequestered in wood products and the biodiversity benefits of a well-managed forest (Reid *et al.*, 2004a).

Biodiversity-related agreements must also incorporate climate objectives. For example, many projects and policies to conserve and sustainably manage ecosystems undertaken by parties to the CBD and other biodiversity-related agreements (such as the Convention on Migratory Species, Convention on Wetlands and

World Heritage Convention) could impact climate change mitigation and adaptation objectives (Secretariat of the CBD, 2003).

Regional policymaking also needs improved integration of climate change and biodiversity concerns. Europe has various policies on each, and although suggestions have been made on how to integrate these concerns (Usher, 2005), policy coherence on these issues remains inadequate. For example, Dudley, (2001) states that the Common Agricultural Policy needs reforming to support the protection of natural woodland in the UK in order that it can adapt to climate change. European forest policy has made some efforts at integrating climate and biodiversity concerns. The Pan European Criteria, Indicators and Operational Level Guidelines for Sustainable Forest Management, adopted in 1998, aim to support carbon sequestration in conjunction with biodiversity conservation. For example they promote afforestation and reforestation with native species. The General Guidelines for the Sustainable Management of Forests in Europe under the Ministerial Conference on the Protection of Forests in Europe does likewise (IUCN, 2004).

Hannah *et al.*, (2002) point out that species range shifts will not respect political boundaries, so effective conservation management will require new regional collaboration. The Pan European Biological and Landscape Diversity Strategy, endorsed in 1995 by 54 European states, recognises the importance of ecological networks and calls for the development of a Pan European Ecological Network (Nijhoff, 2005). The Habitat Directive of the European Union also acknowledges the importance of landscape elements that enhance connectivity, and other global and European policies such as the Bonn and Bern Conventions oblige parties to conserve and manage listed species and habitats (Nijhoff, 2005). Whilst not overtly considering climate change issues, such policies help reduce the vulnerability of biodiversity to climate change.

Nationally, policies and activities that benefit biodiversity and climate change adaptation and mitigation need promoting and mainstreaming into various areas of national policy making (IUCN, 2004). This will avoid a purely sectoral approach and help enhance coordination between different government agencies (IUCN, 2004). But it is not easy as many climate change and biodiversity-related activities are located within the Ministries of Environment, which are traditionally relatively weak when compared to ministries dealing with finance or land use planning (Reid, 2004; Reid *et al.*, 2004b). At present, coordination among sectoral agencies to exploit potential synergies is poor (Secretariat of the CBD, 2003). National plans to deal with climateinduced disasters could identify vulnerable ecosystems as well as vulnerable human settlements (Reid, 2004; Reid *et al.*, 2004b). The development of a workable carbon intensity labelling system, pro-wood building and packaging standards and invigorated recycling programes would help to maximise the climatic advantages of wood use (Reid *et al.*, 2004a). This in turn could provide biodiversity benefits if wood is sourced from well-managed forests. Afforestation and reforestation guidelines could be incorporated into national forest programmes and National Biodiversity Strategies and Action Plans (IUCN, 2004). Protected area policies and other species conservation initiatives also need to become more aware of the effects of climate change and to include this in their prioritisation of hazards and risk management schemes.

Practical Ways to Link Climate Change and Biodiversity

The first step to address climate change must involve reducing greenhouse gas emissions. This requires lifestyle changes to reduce per capita energy consumption, and also technological innovations to increase energy efficiency and help the shift to more renewable technologies. The location of renewable energy initiatives, such as dams or wind farms needs careful consideration of biodiversity impacts. For example, the location of offshore wind farms in Germany is considering ecologically sensitive sites as well as wind energy qualification areas (BMU, 2002). However, climate change is already happening, and will continue to do so whatever steps we now take (IPCC, 2001). We therefore need to look at how we can adapt our countryside to reduce biodiversity losses.

Many conservation practitioners in Europe recommend adopting a landscape approach to reduce the negative impacts of climate change on biodiversity (Hannah *et al.*, 2002). For example, the Woodland Trust in the UK advocates a move from focusing on protecting a few individual sites to a landscape approach. This would involve integrating protection, restoration and extension activities. Wide application of green corridors and buffer areas around particularly important woodland sites is needed (Dudley, 2001). Functionally diverse communities may be better able to adapt to climate change than functionally impoverished systems.

Conserving genotypes and species along with reducing habitat loss, fragmentation and degradation may therefore promote the long-term persistence of ecosystems (Secretariat of the CBD, 2003). In Britain and Ireland, the Modelling Natural Resource Responses to Climate Change (MONARCH) programme of research led by English *Nature* uses modelling approaches to help *nature* conservation policy and management practices adapt to climate change. Results suggest the need for a flexible dynamic approach that can adjust to the changing distribution of species and habitat types (Secretariat of the CBD, 2003). In Europe, the ACCELERATES (Assessing Climate Change Effects on Land use and Ecosystems: from Regional Analysis to The European Scale) Project funded by the European Commission is also assessing the vulnerability of European agroecosystems to environmental change.

Larger protected areas, which cover a range of elevations, microclimates and ecosystems, will be less vulnerable, as species will be able to migrate to another safe habitat within the protected area if climate change adversely affects their present one. Connections between existing protected areas will be increasingly important, as will buffer zones around protected areas, landscape connectivity and management of areas between core protected areas (Hannah *et al.*, 2002; Dudley, 2003; Secretariat of the CBD, 2003). This is the thinking behind Natura 2000: a Europe wide network of protected areas *(http://www.natura2000benefits.org).*

Integrated watershed management can increase water retention and availability in times of drought, decrease the chance of flash floods and maintain vegetation as a carbon sink. It can also conserve watershed biodiversity. Avoiding degradation of peatlands and mires is particularly important for retaining soil carbon stocks and protecting biodiversity (Secretariat of the CBD, 2003).

Certain forest management activities can simultaneously provide biodiversity and climate benefits. These include encouraging native species, increasing rotation age, low intensity harvesting, reduced impact logging, leaving woody debris, harvesting which emulates natural disturbance regimes, avoiding fragmentation, provision of buffer zones, corridors and natural fire regimes, and limiting the use of toxic chemicals and fertilisers (Reid, 2003; Secretariat of the CBD, 2003; IUCN, 2004). Biodiversity and climate benefits are also possible from agroforestry, revegetation, grassland management and agricultural practices. These include conservation tillage, recycling and use of organic materials, maintaining continuous ground cover, intercropping and reduced use of pesticides and herbicides (Secretariat of the CBD, 2003; IUCN, 2004). Planting hedges reduces soil loss and landscape degradation from erosion (problems exacerbated by climate change-induced droughts, floods and winds) thus increasing soil carbon stocks and agricultural productivity, and benefiting biodiversity (Secretariat of the CBD, 2003).

The concept of becoming 'carbon neutral' is gaining popularity with many businesses, which wish to contribute to climate change mitigation activities by offsetting their carbon emissions. Likewise, many nations have committed to reducing their net greenhouse gas emissions under the Kyoto Protocol of the UNFCCC. The CDM provides one mechanism for doing this. Projects designed to sequester carbon, and hence mitigate climate change, present opportunities to incorporate biodiversity considerations. For example, The Netherlands Forest Absorbing Carbon Emissions (FACE) Foundation has invested in forest restoration activities and indigenous tree planting in areas of Mount Elgon National Park, Uganda, that were previously degraded. The Ugandan Wildlife Authority has implemented these activities. The aim is to offset Dutch greenhouse gas emissions from the foundation's clients, which include power generating companies and industrial and business clients in Europe. Certified Emissions Reduction credits would be awarded under the CDM. Tree Farms, a private Norwegian company, has also funded tree planting in the Bukaleba Forest Reserve, Uganda, in anticipation of the CDM becoming operational. Both these projects could have been improved if local community needs had been accounted for (Secretariat of the CBD, 2003).

Using more wood products can sequester significant amounts of carbon, particularly if products have a long useful life and are recycled when no longer useful. If wood is sourced from well-managed forests, biodiversity benefits can also accrue. This is particularly true for European countries where over 90% of imports of roundwood and sawn wood are from other European states and most forests are managed sustainably (Reid *et al.*, 2004a). For short life cycle packaging materials, substituting one tonne of virgin card for glass, plastic, steel or aluminium results in average savings of 1.1, 2.8, 2.9 and 4.1 tonnes of carbon dioxide respectively (Ministry for Environment, 2001). Despite this, many people think using wood substitutes is better for the environment (Reid *et al.*, 2004a).

Several possible tools for integrating biodiversity and climate change concerns exist. The ecosystem approach could incorporate climate concerns, and Environmental Impact Assessments, and Strategic Environmental Assessments can be adapted to support broad uptake of environmental priorities. For example, Finland is applying a Strategic Environmental Assessment approach in developing its national climate strategy. Using these approaches to assess climate adaptation and mitigation projects, however, might raise assessment and compliance costs and ultimately prevent beneficial projects from occurring (Secretariat of the CBD 2003).

Conclusions

The impacts of climate change on European biodiversity can already be seen in terms of changes in species' ranges and ecosystem boundaries, shifts in reproductive cycles and growing seasons, and changes to the complex ways in which species interact (predation, pollination, competition and disease). The extent of these effects varies between regions, and between species and ecosystems, some of which are more vulnerable than others. Existing protected areas are unlikely to be particularly effective in the face of expected changes, and many species extinctions are likely.

Strategies adopted to mitigate or adapt to climate change, such as construction of sea walls, dams and wind farms, or afforestation and reforestation activities, also impact biodiversity. And land use changes (such as cutting down forests or draining peatlands) and subsequent changes to biodiversity can also alter levels of greenhouse gas emissions thus affecting the global climate. The many linkages between climate and biodiversity suggest the need for more integrated policy responses, at international, regional and national levels. Nationally, activities that meet biodiversity and climate change objectives need promoting and mainstreaming into various sectoral areas of policy making. Reducing energy consumption, increasing energy efficiency and promoting renewable energy technologies are priorities. But existing trends in temperature changes, precipitation level changes and sea level rise will continue regardless. The unprecedented speed of this change provides much cause for concern. A continuing increase in the frequency of extreme climatic events may also affect biodiversity more than the gradual changes expected.

Activities which meet both climate and biodiversity objectives therefore need exploring. These include adopting new landscape management approaches, and ensuring carbon sequestration projects and renewable energy projects incorporate biodiversity considerations.

Such approaches are also relevant to a broader audience than Europe. Indeed, rapidly expanding human populations in countries like India also make conservation strategies that exclude local people from relatively small protected areas both inequitable and inadequate in the context of climate change. Just as extensive landscape management approaches are needed in Europe, broader inclusive conservation strategies involving the wider landscape and seascape are appropriate and necessary in low-income countries to effectively tackle large global issues such as climate change.

Acknowledgements

This paper was first presented at Biodiversity Loss in Europe: Symposium of the Spanish Terrestrial Ecology Association and the European Ecological Federation, 4-6 March 2005. The author would like to thank Kim Jensen, Kartik Shanker and Saleemul Huq for their help and advice in preparing this paper.

(Hannah Reid, Research Associate, International Institute for Environment and Development, 3 Endsleigh Street, London WC1H 0DD, UK. The author can be reached at Hannah.reid@iied.org).

References

Bakkenes M, J R M Alkemade, F Ihle, R Leemans and JB Latour 2002, Assessing effects of forecasted *climate change* on the diversity and distribution of European higher plants for 2050. *Global Change Biology* 8:390-407.

Beaugrand G, F Ibañez, J A Lindley and P C Reid 2002, Diversity of calanoid copepods in the North Atlantic and adjacent seas: Species associations and biogeography. *Marine Ecology Progress Series* 232:179-195.

Beebee T J C 2002, Amphibian phenology and *climate change. Conservation Biology* 16:14540-1455.

BMU 2002, *Strategy of the German Government on the Use of Off-Shore Wind Energy.* German Government, Berlin, Germany.

Broadmeadow M 2000, *Climate Change–Implications for Forestry in Britain.* Forestry Commission, Edinburgh, United Kingdom.

Brown S, J Sathaye M Cannell and P E Kauppi 1996, Management of Forests for Mitigation of Greenhouse Gas Emissions. In: *Climate Change 1995–Impacts, Adaptations and Mitigation of Climate Change: Scientific-Technical Analyses* (eds. R T Watson, MC Zinyowera, RH Moss and D J Dokken), pp. 773-797. Contribution of Working Group II to the Second Assessment Report of the Intergovernmental Panel on *Climate Change*, Cambridge University Press, Cambridge, United Kingdom.

Chuine I and E G Beaubien 2001, Phenology is a major determinant of tree species range. *Ecology Letters* 4:500-510.

Clark R A, C J Fox, D Viner and M Livermore 2003, North Sea cod and climate change–modelling the effects of temperature on population dynamics. *Global Change Biology* 9:1669-1680.

Crick HQ P, C Dudley, D E Glue and D L Thomson 1997, UK birds are laying eggs earlier. *Nature* 388:526.

Dudley N 2001, *A Midsummer Night's Nightmare? The Future of UK Woodland in the Face of Climate Change.* The Woodland Trust, Grantham, United Kingdom.

Dudley N, 2003, No Place to Hide: *Effects of Climate Change on Protected areas.* WWF *Climate Change* Programme, Godalming, United Kingdom.

EEA 2004, *Impacts of Europe's Changing Climate. An Indicator-Based Assessment.* European Environment Agency, Copenhagen, Denmark.

Fearnside P M 2001, Environmental impacts of Brazil's Tucurui Dam: Unlearned lessons for hydroelectric development in Amazonia. *Environmental Management* 27(3):377-396.

Fu C Z, JH Wu, J K Chen, QH Qu and G C Lei 2003, Freshwater fish biodiversity in the Yangtze River basin of China: Patterns, threats and conservation. *Biodiversity and Conservation* 12(8):1649-1685.

Hannah L, G F Midgley, T Lovejoy, W J Bond, M Bush, J C Lovett, D Scott and F I Woodward 2002, Conservation of biodiversity in a changing climate. *Conservation Biology* 16(1):264-268.

Houghton J T, Y Ding, D J Griggs, M Noguer, P J van der Linden, X Dai, K Maskell and C A Johnson (eds.). 2001. *Climate Change 2001–The Scientific Basis.* Contribution of Working Group I to the Third Assessment Report of the Intergovernmental Panel on *Climate Change.* Cambridge University Press, Cambridge, United Kingdom.

Hughes L 2000. Biological consequences of global warming: Is the signal already apparent? *TREE* 15(2):56-61.

Huq S and H Reid 2005, Benefit Sharing under the Clean Development Mechanism. In: *Making Kyoto Work: Legal Aspects of Implementing the Kyoto Protocol Mechanism* (eds. D Freestone and C Streck), pp. 229-247. Oxford University Press, Oxford, United Kingdom.

IPCC 2001, *Climate Change 2001: Impacts, Adaptation, and Vulnerability.* Contribution of Working Group II to the Third Assessment Report of the Intergovernmental Panel on *Climate Change* Published for the Intergovernmental Panel on *Climate Change* by Cambridge University Press, Cambridge, UK. 1005 pages.

IUCN 2004, *Afforestation and Reforestation for Climate Change Mitigation: Potentials for Pan-European Action.* IUCN – The World Conservation Union – Programme Office for Central Europe, Gland, Switzerland.

Janssens I A, A Freibauer, P Ciais, P Smith, G J Nabuurs, G Folberth, B Schlamadinger, R W Hutjes, R Ceulemans, E D Schulze, R Valentini and A J Dolman 2003, Europe's terrestrial biosphere absorbs 7 to 12% of European anthropogenic CO2 emissions. *Science Express* 300(5623):1538-1542.

Kappelle M, M M I Van Vuuren and P Baas, 1999. Effects of climate change on biodiversity: A review and identification of key research issues. *Biodiversity and Conservation* 8: 1383-1397.

Klein Tank, A M G 2004. Changing temperatures and precipitation extremes in Europe's climate of the 20th century. Ph.D. thesis. Utrecht University, Utrecht, The Netherlands. 124pp.

Lanchbery J 2004, *Ecosystem Loss and its Implications for Greenhouse Gas Concentration Stabilisation.* Paper presented at the International Symposium on Stabilisation of Greenhouse Gases, 1-3 February 2005, Met Office, Exeter, United Kingdom.

Leemans R and A van Vliet 2004, *Extreme Weather: Does Nature Keep Up?* WWF/Wageningen University, The Netherlands.

Lemoine N and K Böhning-Gaese 2003, Potential impact of global climate change on species richness of long-distance migrants. *Conservation Biology* 17(2):577-586.

Liebsch G, K Novotny and R Dietrich 2002, *Untersuchung von Pegelreihen zur Bestimmung der Änderung des mittleren Meeresspiegels an den europäischen Küsten.* Technische Universität Dresden (TUD), Germany.

Menzel A and P Fabian 1999, Growing season extended in Europe. *Nature* 397:659.

Ministry for Environment 2001, *Environmental Impact of Packaging Materials.* Ministry for Environment, Denmark.

Moen J, K Aune, L Edenius and A Angerbjörn 2004, Potential effects of climate change on treeline position in the Swedish mountains. *Ecology and Society* 9(1):16.

Molau U and J M Alatalo 1998, Responses of sub-alpine plant communities to simulated environmental change. *Ambio* 27:332-329.

Mortished C 2005, Post-Katrina US can learn plenty from the Dutch. Europe infrastructure spending. *Globeandmail.com,* Thursday September 8th 2005, Canada. *http: // www.globetechnology.com/servlet/ArticleNews/TPStory/LAC/20050908/IBEUROPE08/ TPTechInvestor/*

Nabuurs G-J, A Pussinen, T Karjalainen, M Erhard and K Kramer 2002, Stemwood volume increment changes in European forests due to climate change – a simulation study with the EFISCEN model. *Global Change Biology* 8:304-316.

Nijhoff d P 2005, Climate Change and Biodiversity: Challenges for European Policy. *http://www.rlg.nl/website/uitgaven/lezingen/lezing_14.html*

Page SE, F Siegert, J O Rieley, HD Boehm, A Jaya and S Limin. 2002. The amount of carbon released from peat and forest fires in Indonesia during 1997. *Nature* 420:61-65.

Parmesan C, N Ryrholm, C Steganescu J K Hill, C D Thomas, H Descimon, B Huntley, L Kaila, J Kullberg, T Tammaru, W J Tennent, J A Thomas and M Warren 1999, Poleward shifts in geographical ranges of butterfly species associated with regional warming. *Nature* 399:579-583.

Parmesan C and G Yohe 2003, A globally coherent fingerprint of climate change impacts across natural systems. *Nature* 421:37-42.

Pauli H, M Gottfried and G Grabherr 2001, High Summits of the Alps in a Changing Climate. The Oldest Observation Series on High Mountain Plant Diversity in Europe. In: *Fingerprints of Climate Change – Adapted Behaviour and Shifting Species Ranges* (eds. G R Walther, C A Burga and P J Edwards), pp. 139-149. Kluwer Academic Publisher, New York, United States of America.

Pinol J, J Terradas and F Lloret 1998, Climate warming, wildfire hazard, and wildfire occurrence in coastal eastern Spain. *Climatic Change* 38:345-357.

Post E, N C Stenseth, R Langvatn and J M Fromentin 1997, Global climate change and phenotypic variation among red deer cohorts. *Proceedings of the Royal Society of London, Series B,* 264:1317-1324.

Pounds J A, M P L Fogden and J H Campbell 1999, Biological response to climate change on a tropical mountain. *Nature* 398:611-615.

Reid H 2003, A framework for biodiversity and climate. *Tiempo* 50:7-10.

Reid H 2004, Climate Change – Biodiversity and Livelihood Impacts. In: *The Millennium Development Goals and Conservation* (ed. D Roe), pp. 37-54. IIED, London, United Kingdom.

Reid H, S Huq, J MacGregor, D Macqueen, J Mayers, L Murray, R Tipper and A Inkinen. 2004a. *Using Wood Products to Mitigate Climate Change: A Review of Evidence and Key Issues for Sustainable Development*. IIED, London, United Kingdom.

Reid H, B Pisupati and H Baulch 2004b, How Biodiversity and Climate Change Interact. SciDev.Net Biodiversity Dossier Policy Brief. *http://www.scidev.net/dossiers/index.cfm?fuseaction =policybriefs&dossier=11*

Root T L, J T Price, K R Hall, S H Schneider, C Rosenzweig and J A Pounds 2003, Fingerprints of global warming on wild animals and plants. *Nature* 421:57-60.

RSPB 2004a, *Climate Change and Birds*. RSPB information leaflet, Sandy Bedfordshire, United Kingdom.

RSPB 2004b, *Wind Farms and Birds*. RSPB information leaflet, Sandy Bedfordshire, United Kingdom.

Sciare J, N Mihalopoulos and F J Dentener 2000, Inter-annual variability of atmospheric dimethylsulfide in the southern Indian Ocean. *Journal of Geophysical Research*. 105:26,369-26,377.

Secretariat of the CBD 2003, *Interlinkages Between Biological Diversity and Climate Change: Advice on the Integration of Biodiversity Considerations into the Implementation of the United Nations Framework Convention on Climate Change and Its Kyoto Protocol*. CBD Technical Series No. 10. Secretariat of the CBD, Montreal, Canada.

Tamis W L M, M Van't Zelfde and R Van der Meijden 2001, Changes in Vascular Plant Biodiversity in the Netherlands in the 20th Century Explained by Climatic and other Environmental Characteristics. In: *Long-term Effects of Climate Change on Biodiversity and Ecosystem Processes* (eds. H Van Oene, W N Ellis, M M P D Heijmans, D Mauquoy, W L M Tamis, F Berendse, B Van Geel, R Van der Meijden and S A Ulenberg), pp. 23-51. NOP, Bilthoven, The Netherlands.

Theurillat J-P and A Guisan 2001, Potential impact of climate change on vegetation in the European Alps: A review. *Climatic Change* 50:77-109.

Thomas C D, A Cameron, R E Green, M Bakkenes, L J Beaumont, Y C Collingham, B F Erasmus, M F De Siqueira, A Grainger, L Hannah, L Hughes, B Huntley, A S Van Jaarsveld, G F Midgley, L Miles, M A Ortega-Huerta, A T Peterson, O L Phillips and S E Williams 2004, Extinction risk from climate change. *Nature* 427:145-148.

Thomas, C D and J J Lennon 1999, Birds extend their ranges northwards. *Nature* 399:213.

Usher M B 2005, *Conserving European Biodiversity in the Context of Climate Change*. Committee for the activities of the Council of Europe in the field of biological and landscape diversity. Council of Europe, Strasbourg.

Virtanen T and S Neuvonen 1999, Climate change and macrolepidopteran biodiversity in Finland. *Chemosphere: Global Change Science* 1:439-448.

Walther G-R, E Post, P Convey, A Menzel, C Parmesan, T Beebee, J-M Fromentin, O Hoegh-Guldberg and F Bairlain 2002, Ecological responses to recent climate change. *Nature* 416:389-395.

World Commission on Dams 2000, *Dams and Development: A New Framework for Decision Making*. Earthscan, London, United Kingdom.

Index

A

B

G

H

I

K

L

M

N

Q

R

S